ECONOMICS OF RIVER FLOWS

ECONOMICS OF RIVER FLOWS
Lessons From Dam Removals in America

Editor
BHARAT JHUNJHUNWALA

KALPAZ

Economics of River Flows

Rs. 750

ISBN: 978-81-7835-816-1

Published in 2010 in India by
Kalpaz Publications
C-30, Satyawati Nagar,
Delhi-110052
E-mail: kalpaz@hotmail.com
Phone : 9212729499

Laser Type Setting by: Rudra Computer Graphics, Delhi
Printed at : Chawla Offset Printers, Delhi

Contents

Preface

This book of readings has its origins in my involvement in opposing the construction of hydropower dams on the Ganga at Dev Prayag. As a trained economist it was a matter of great curiosity for me that dams were being made in India while they were being removed in the United States—both for securing economic growth. Deeper study indicated that the difference lay in the fact that many costs of hydropower were ignored in India while benefits were overstated. Dams are being built in India because economic analysis is flawed and ignores various costs that are surreptitiously passed on to the society.

I had an opportunity to visit Tunisia and the United States to visit my daughters in 2008. I used this opportunity to study the working of dams in Tunisia and removals in the United States. This volume is a result of that humble effort.

I hope our planners will look at the lessons of dam removals in the United States.

Bharat Jhunjhunwala
Lakshmoli, PO Maletha, Kirti Nagar
Uttarakhand 249161 India

Acknowledgements

This collection of readings has been made possible by a large number of friends and river lovers.

Don W Baldwin of Susquehanna Electric showed me around the Conowingo Dam.

Mark Chernaik of Environmental Law Alliance Worldwide, Carl Bruch of Environmental Law Institute and Shannon Wilson of Seirra Club provided me with key contacts.

Jock Conyngham has pointed out the limits of Cost-Benefit Analysis and drawn attention to the fact that some values are absolute.

Geoffrey M Goll and John A. Miller of Princeton Hydro explained issues relating to valuation of natural capital.

Fethi Lebdi of Ministry of Irrigation, Government of Tunisia and Yadh Zahar, Professor, University of Tunis explained the water supply system in that country. Nabil Karray showed the coastal erosion sites in Tunisia.

Stephanie Lindloff assisted me in connecting with other activists of American Rivers and also provided with some of the papers included in this volume. Sara Deuling, along with Vince Humenay of Pennsylvania Department of Environment, showed me the site of the removed Ralph Stover Dam.

Jim Puzzuto of University of Delaware educated me about the role of sediments in water quality.

Maya K. van Rossum of the Delaware Riverkeepers has pointed out that dams for flood control are still being built in the United States. Chari Towne of the same organization guided me to visit the Angelica and Felix dams.

John Seebach of American Rivers provided me with many of the documents included in this volume. He explained the issues relating to valuation of electricity.

Electricity consultant Shankar Sharma has given me confidence to publish this book by endorsing the importance of Cost-Benefit Analysis.

Himanshu Thakkar of Dams, Rivers and People initially exposed me to the fact that dams were being removed in large numbers in the United States.

John P. Tippet of Friends of Rappahannock showed me the site of the removed Embrey Dam and also explained the working of the side stream skimmer at the Motts Run Reservoir.

Paritosh Tyagi has raised many questions on an earlier draft of the manuscript. That has helped this writer dwell deeper into many issues.

Hal Wiggins of U.S. Army Corps of Engineers kindly provided with a copy of his book *A Tale of Two Dams.*

The authors and sponsoring organizations of the papers included in this book have been kind to allow use of the articles in this volume.

I deeply acknowledge the great value of education received at the Department of Food and Resource Economics, University of Florida. My small ability to grasp and deal with the economics presented in this volume is wholly learnt here.

My daughters Juhi and Radhika and son-in-law Varun hosted me and Madhu during the time we studied dams in Tunisia and United States. Thank you all!

Abbreviations

ACOE	Army Corps of Engineers
aMW	Average Megawatts
AF	acre-feet
AMC	Amorphous Metal Core
ANB	Annual Net Benefits
ARNI	Aquatic Resources of National Importance
BCA	Benefit-cost analysis
BEA	Bureau of Economic Analysis
BLM	Bureau of Land Management
BO	Biological Opinion
BPA	Bonneville Power Administration
CBA	Cost-Benefit Analysis
CC	Combined Cycle (Plant)
CEA	Cost Effectiveness Analysis

CEA	Central Electricity Authority
CEC	California Energy Commission
CEQ	Council on Environmental Quality, Executive Office of the President
CERP	Comprehensive Everglades Restoration Plan
CFS	Cubic Feet per second
COE	U.S. Army Corps of Engineers
CT	Combustion Turbine (Plant)
CV	Contingent Valuation
CVM	Contingent Valuation Method
Corps	U.S. Army Corps of Engineers
CPUC	California Public Utility Commission
DEIS	Draft Environmental Impact Statement
DEP	Connecticut Department of Environmental Protection
DNR	Minnesota Department of Natural Resources
DOE	Department of Energy
DOI	Department of the Interior
EA	Economic Impact Analysis
ECPA	Electric Consumers Protection Act
EIA	Economic Impact Analysis
EIS	Environmental Impact Statement
EPA	Environmental Protection Agency
ESA	Endangered Species Act

FERC	Federal Energy Regulatory Commission
FOR	Friends of Rappahannock
FPA	Federal Power Act
FWS	Fish and Wildlife Service
GAO	General Accounting Office
gpm	gallons per minute
HEA	Habitat Equivalency Analysis
HQUSACE	Head Quarters of U.S. Army Corps of Engineers
HRC	Hydropower Reform Coalition
IPCC	Intergovernmental Panel on Climate Change
KIP	Klamath Irrigation Project
MBF	Million Board Feet
mgd	Million gallons per day
MOA	Memorandum of Agreement
MWRA	Massachusetts Water Resources Authority
NEPA	National Environmental Policy Act
NHPA	National Historic Preservation Act
NOAA	National Oceanic and Atmospheric Administration
NPDES	National Pollutant Discharge Elimination System
NPS	National Park Service
NRCS	Natural Resources Conservation Service
NRRSS	National River Restoration Science Synthesis
NWPP	Northwest Power Pool

O&M	Operation and Maintenance
ODOT	Oregon Department of Transportation
OMB	Whitehouse Office of Management and Budget
ORV	Outstanding and remarkable value
PFO	Palustrine Forested Wetlands
PUC	Public Utility Commission
RIA	Regulatory Impact Analysis
RIMS	FERC Records and Information Management System
RIMS	Regional Input-Output Modeling System Bureau
RMA	Resource Management Act
RMP	River Management Plan
ROR	Run-of-River
RORP	Run-of-River with Pondage
RVM	Replacement Value Method
SMURRF	Santa Monica Urban Runoff Recycling Facility
SRAFRC	Coalition of various agencies on Susquehanna River Basin
T&D	Transmission and Distribution
T&E	Threatened and Endangered
TEV	Total Economic Value
TMDL	Total Maximum Daily Load
UNEP	United Nations Environment Programme
USBR	US Bureau of Reclamation

USACE	United States Army Corps of Engineers
USDA	Department of Agriculture
USGS	U.S. Geological Survey
USFWS	U.S. Fish and Wildlife Service
VDGID	Commonwealth of Virginia
VEPCO	Virginia Electric and Power Company
WBIS	Wilkes-Barre Inflatable Structure
WRC	Water Resources Council
WSRA	Wild and Scenic Rivers Act of 1968
WTA	Willingness to Accept
WTP	Willingness to Pay or Water Treatment Plant
WVLRP	Wyoming Valley Levee Raising Project

List of Contributors

Jock Conyngham, J. Craig Fischenich and Kathleen D. White, Environmental Laboratory, Cold Regions Research and Engineering Laboratory, Engineer Research and Development Center, U.S. Army Corps of Engineers.

Betsy Otto, American Rivers.

Taxpayers for Common Sense and others, REP, Pacific Coast Association of Fishermen's Associations, Northwest Sportfishing Industry Association, NW Energy Coalition, Institute of Fisheries Resources, American Rivers

Michael Paulson, Seattle Post-Intelligencer, Washington Correspondent

Hal Wiggins, US Army Corps of Engineers (USACE)

David Marcus, Hydropower Reform Coalition.

Jon H. Goldstein, For American Rivers.

Pamela Baldwin, Legislative Attorney, American Law Division.

John Loomis, Professor, Department of Agricultural and Resource Economics, Colorado State University.

Ann L. Riley, Watershed and River Restoration Advisor, San Francisco Bay Regional Water Quality Board, Oakland, Calif, Jan 2008.

Introduction: Lessons from Dam Removals in the United States

Bharat Jhunjhunwala

Dam are being removed in the United States in large numbers while they are being made in equally large numbers in India. Both are being done to promote economic development. The price of electricity is also similar in the global bazaar. The methodology of Cost-Benefit Analysis (CBA) followed by both countries is also same. Yet the results obtained are diametrically opposite. The United States finds many dams to be uneconomic while India mostly finds them to be economic. This happens because the Cost-Benefit Analysis undertaken by us in India ignores many costs and overestimates the benefits.

However, there are crucial differences between the situations in the two countries. Dams being removed in the United States have mostly become obsolete from the economic standpoint. Many were made to facilitate movement of barges over steep flows of rivers. This has almost wholly been replaced by road transport. The Felix Dam near Reading, Pennsylvania was made for this purpose. It was not in use for many years when it got breached in 1999. There was no economic benefit of rebuilding it because it was no longer used for navigation. On the other hand, it had become a safety hazard. People would walk to the edge of the broken structure and fall and injure themselves. Removal of the dam did not impose any noticeable costs on the economy. The community living nearby nevertheless wanted to rebuild the dam. They feared that removing the dam and the upstream impoundment would lead to decline in the value of their property. This writer was told by a property owner near the dam that "People were knocking at the door seeking support for rebuilding the dam." But the Government refused to put in the money to rebuild the dam and instead paid for removing it. The benefits from increase in property prices were reckoned to be much less than the cost of rebuilding the structure.

Fig. I : Disused Lock at Decommissioned Felix Dam

A similar situation prevailed at the nearby Angelica Dam. This dam was breached by floods about 15 years ago. The impoundment was earlier a picnic destination for the local people. There was fishing and boating. There was a restaurant teeming with visitors. Local people wanted to rebuild the dam. One person who lived in his camper on the site told this writer: "Previously this place was full of visitors and anglers. It was nicer when there was water all around. The pavilion had a restaurant and row boats and canoes. People wanted to rebuild the dam. They did dances to raise money for rebuilding it." But the cost was high and the City did not find the money to do so. It was removed.

Fig. II : This is where the Angelica Dam once stood

The Ralph Stover Dam, Roger Park was built about 150 years ago to provide water for running a mill. The impoundment was converted for use as swimming pool in early 1900s. It was removed due to safety considerations even though local people wanted to retain it for swimming.

These are typical examples of dams that have been removed in large numbers in the United States. Almost all the dams being removed are small run-of-river structures which produced little or no electricity. Thus, the experience of decommissioning of dams in the U.S. is not directly applicable

Fig. III : Mouth of the disused channel taking water from the Ralph Stover Dam to the mill

to us. Yet the experience is relevant for us because the underlying method of benefit-cost analysis adopted by the U.S. is applicable to our situation as well. The dams were removed because they no longer provided mill- or hydro power and the benefits from recreation were assessed to be less than the cost of remaking them (Please see *Chapter 3: Economic Analysis of Removing Four Dams on the Lower Snake River*).

At the same time, few live dams have also been removed. The Edwards Dam in Maine was removed even though it was producing electricity because the cost of loss of fisheries was reckoned to be more than gains from generation of electricity (Please see *Chapter 2: Funding Sources for Dam Removal*).

It must be mentioned that some new dams are also being made in the United States while others are being removed. New dams are being made mostly for providing drinking water to the cities and for flood control. The economic value of drinking water is very high hence dams made for this objective are likely to stand economic scrutiny. The economic benefits of flood control dams are probably less clear. In any event, very few dams, if any, are being constructed for generation of hydropower. Thus, as far as hydropower dams are concerned, the position in India and the United States is exactly opposite.

The basic point is that *in certain conditions* economics favours removal of dams. Dams are not always economically beneficial.

The more important lesson is that changing economic conditions have lead to the dams becoming uneconomic. They become safety hazards with passage of time. Our needs of hydropower may similarly change in future (Please see *Chapter 2: Funding Sources for Dam Removal*).

Dams have much contributed to economic development of the United States earlier. However, change in economic circumstances led to dams becoming economically unviable. Construction of large nuclear plants led to low price of electricity and rendered small hydropower plants like Embrey Dam unviable. The hydropower plants being built by us may get similarly outdated with cheap solar power and mega thermal power plants in the making (Please see *Chapter 6: Removal of Embrey Dam*).

The Elhwa Dam was removed, among others, because one industrial water customer closed down and because new technologies made it possible to recycle concrete from the dam (Please see *Chapter 4: Non-use Values lead to the Removal of Elwha Dam*).

Change of values in the American society has also contributed to this reassessment. Previously the major focus was on economic development. Nowadays there is more value attached to ecology and environment. In the last three decades, growing concern over environmental quality, mounting flood losses, endangered species and aesthetic characteristics of landscapes have become more prominent in the national discourse. It has taken decades for some environmental changes caused by the larger structures to become apparent. This is leading to removal of dams in the United States.

Economic development involves a progression. Initially the focus is on the satisfaction of material desires which would include consumption of electricity for washing machines etc. However, at higher stages, the social and spiritual needs become more important. These are subsumed under the category 'environment'. This progression from material- to spiritual desires has led to the removal of dams in the U.S. We in India are likely to soon go through a similar progression. Our needs for electricity will reduce while our needs for social and spiritual betterment will become more prominent. Therefore, we must factor in the sacrifice of long term social and spiritual objectives while taking a decision to make a dam for the provision of electricity in the short term. Our ability lies in satisfying our short-term lower material needs without sacrificing the long term spiritual needs.

We in India are caught between two opposite forces. There is a great commitment to secure economic development so that we never again have to go through the horrible experience of foreign domination. On the other hand, we worship our rivers. This latter value is suppressed at the present as we are solely preoccupied with economic development. But this will pass. The economic benefits from dams will recede in the background in India just as they have done in the United States. There is great possibility that we will attach more value to higher pursuits of life such as mental peace obtained from sitting on river banks and less value to the consumption of power. Therefore, there is a need to be careful in making long term commitment of our rivers for generation of hydropower (Please see *Chapter 4: Non-use Values lead to the Removal of Elwha Dam*).

Changes in the structure of our economy may make the dams useless. For example, solar power may be available in near future at a fraction of present cost. That will make hydropower dams a loss proposition. The development of solar power at Rs 4.00 per kWH, as proposed by some players, would render many hydropower projects uneconomic. Or development of LED bulbs can reduce peak demand. The flow in our rivers may reduce due to recession of glaciers. The least our planners can do is to provide a statement of such uncertainties and how they will affect the project. We should not commit our rivers for a long period for short term gains. A sensitivity analysis of future power scenario is a must before building dams (Please see *Chapter 9: Cost-Benefit Analysis*).

The possible change in the economic scenario is sometimes used to support hydropower. Argument is that it is beneficial to produce hydropower in *present* economic conditions. The dams can be dismantled when and if they become uneconomic. This logic is correct. But it is necessary to make true assessment of costs and benefits both in the short- and long run. Then policymakers may indeed decide to harness the short term gains while foregoing the long term benefits. But it is essential that an informed decision may be taken on such tradeoffs. Such informed decision is only possible if a true cost-benefit analysis is done.

Government Agencies must lead Decommissioning

A range of Government and non-Government agencies are involved in dam removal in the United States. These agencies have determined that their objectives will be furthered by funding dam removal.

Our Ministry of Environment and Forests is parallel to the U.S. Environment Protection Agency. The EPA used its funds for Chesapeake Bay Program to pay for dam removal to allow fish migration.

Our State Forest Departments have the mandate to preserve wildlife. This is parallel to the Pennsylvania Fish and Boat Commission which has provided funds for dam removal.

The courts often impose penalties upon units making water and air pollution. These monies should be used for removal of or prevention and construction of new dams since the dams will add to pollution. The Connecticut Department of Environment Protection has done precisely this.

State Governments in the United States have River Restoration Funds. This is parallel to our Ganga Action Plan. These funds must be used to remove barrages that cause river destruction.

Our State Departments of Fisheries and Wildlife have the mandate to develop fish resources. The National Fish and Wildlife Foundation of the U.S. has a similar mandate. This agency is funding dam removal because it helps develop fish resources. Such funding must also be provided by Ministry of Tourism because free flow of river waters promotes tourism (Please see *Chapter 2: Funding Sources for Dam Removal*).

Hydropower is not renewable

The common belief in India is that Hydropower is 'renewable' while thermal power is exhaustible. Experience of the United States indicates otherwise. The life of dams is reckoned to be about 50-70 years. After this period they require extensive repairs. Dams develop cracks in course of time. The Embrey Dam was constructed in 1910. It developed cracks and stared leaking and became unsafe in 1990s. Lifespan of dams being built by us will likewise be a few decades. The water resource used for generation of hydropower may be renewable but the dam or the site may not be 'renewable' (Please see *Chapter 6: Removal of Embrey Dam*).

Two types of Run-of-River (ROR) hydropower dams are being made in India. A small barrage is built and water is diverted into a tunnel in 'pure' RORP dams. Water is taken through the tunnel to the power station where it is let into the river again. The gradient between the barrage and power station is used to generate hydropower. The tunnel may be as long as 30 km. Such barrages create a small impoundment of water at the barrage. These barrages may be renewable. They could be taken out easily if damaged and rebuilt at the same site.

The second type dams are 'Run of River with Pondage' (RORP). A dam of about 100 meters height is built on the river. A large reservoir extending backward to about 30 km is created. The gradient between the top and bottom of the dam structure is use to generate hydropower. Such RORP dams are less renewable. Large amount of sediment is trapped behind the dam. Arrangements for flushing the trapped sediments are made in RORP dams. However, flushing only removes sediments deposited in the one- or two kilometers behind the dam. Flushing does not remove sediments trapped father behind. It becomes difficult to remove and rebuild a dam because these sediments will flow as soon as the dam structure is removed. Removal of the structure allows this huge amount of sediment to flow and causes immense destruction downstream. Cost of removal of sediment becomes exorbitant. In the result an obsolete dam has to be maintained simply to prevent flow of sediments.

The deterioration of dam structures is a significant cost in the long run. Dam companies account for maintenance costs for 30-year life on which their economic viability is worked out. The costs to be incurred after this period are left unaccounted. The lease agreements with dam companies that this author has seen do not provide for dam removal costs. Implicitly these costs are to be borne by the society when the dams will be decommissioned while the benefits are reaped by the dam companies at present (Please see *Chapter 1: Problems of Dam Removal*).

There is also a 'scarcity value' of natural resources. The value of a particular resource rises as it becomes scarce. Once a cascade of dams is built on our rivers, it leaves less free flow remaining that can be used for generation of yet more hydropower. The value of the remaining free flow increases as the availability reduces. This 'depletion premium' is not accounted in cost calculations (Please see *Chapter 18: Economic Value of New Jersey Rivers*).

Comprehensive evaluation of economics and environmental impacts

The approval of hydropower dams in India follows two independent procedures. The main approval is granted by Central Electricity Authority (CEA) which looks at the shortage of power in the country and price at which it will be produced by a proposed project. The CEA does not look at the environmental aspects of a proposed project. However, it builds-in a condition that clearance will be obtained from Ministry of Environment and Forests (MOEF). In doing so, the CEA assumes that the conditions imposed by the MOEF will not impact the cost of production from the project.

The project developer approaches the MOEF for environment clearance after obtaining clearance from CEA. MOEF often recognizes that a dam will impose certain environmental harm. But it does not stop the project because it *assumes* that the benefits from generation of electricity are large and the cost to the environment is less. It does not even impose conditions such as installation of fish elevators because that would impose large cost upon the project developer. MOEF only imposes small conditions like seeding of fish in the reservoir to keep the costs imposed upon the developer low. Neither CEA nor MOEF makes a comprehensive assessment of economic and environmental costs and benefits together. CEA looks as the economic aspect and clears the project. MOEF assumes that economic benefits are large and allows imposition of environmental costs for reaping that unspecified benefit. In the end, projects that are harmful for the economy due to environmental impacts may be cleared because no comprehensive assessment of economic and environmental costs is done.

The United States follows a different logic. The Federal Energy Regulatory Commission (FERC) looks at both the economic and environmental aspects together in a comprehensive manner. The FERC itself considers impacts on fish and wildlife habitat, recreational opportunities and environmental quality rather than shifting this burden upon another authority. The underlying concept is that protection of these values raises the cost of production of electricity. For example, minimum flow requirements reduce the production of power and increase its price. These are integral components of project appraisal in the U.S. (Please see *Chapter 11: Critique of FERC's Approval to Klamath Hydroelectric Dam*).

The Edwards dam in Maine was required by the Federal Energy Regulation Commission to install fish ladders to enable salmon and other fish to migrate upstream to their spawning areas. The owner of Edwards Dam, facing a $9 million price to build a fish ladder, agreed to allow the dam to be removed (Please see *Chapter 5: Removal of Edwards Dam*). There is, therefore, a need to establish one combined authority that looks at all aspects of a dam.

Economic Value of Electricity

Hydropower dams are being made in India in large numbers under the assumption that the benefits are large. It is not considered necessary to make a calculation of the benefits. Experience of the United States indicates a need to undertake such exercise. Four types of overestimation of benefits are indicated. One, the demand for electricity in India is overestimated. Two, high prices of electricity prevailing in the short run are used as a justification to produce more electricity in the long run. Three, social benefits arising from the use of electricity are not explicitly calculated. It is assumed they are large, which may not be the case. Four, production of hydropower during the monsoons is valued at the same rate as during winters and summers.

Demand for Electricity

The practice in India is to estimate the prevailing shortage of electricity and then put up generation projects to produce as much power. This approach ignores the fact that present high demand is, in part, due to low price at which electricity is made available to the consumer.

The cost of production increases as the level of production increases. Production of the first unit of electricity may cost Rs 1 per KwH because a favourable site for hydropower project is available. This cost increases as the level of production increases and less favourable sites have to be exploited. This is the supply curve of power.

On the other hand, the benefits decrease with an increase in the level of consumption. The first few units of electricity provided to a household are used for lighting and has a huge value. The subsequent units used for water heating etc. have less value. This is the demand curve of power.

The optimum level of production of power is determined by the intersection of supply and demand. The costs increase while benefits decrease as we raise the level of production. At some point the cost becomes equal to the benefit. This is the optimum level of production.

It follows that it is not beneficial to meet all the shortage of electricity that may exist today. Our planners do not examine this declining benefit from increased level of consumption of electricity. If all shortages were to be removed as aimed by our planners, the benefits would certainly be less than the costs. Such production will be a loss proposition for the economy. Our planners are aiming to 'over-produce' power by seeking to remove all shortages (Please see *Chapter 11: Critique of FERC's Approval to Klamath Hydroelectric Dam*).

High Short Term Price of Electricity

A market for electricity is already in operation in India. According to one news report the average cost of power that was traded was Rs. 4.50 per unit in 2007-08. Power is available at this price. This price relates to short-term requirements or spot market. Most power in the country, however, is traded on the basis of long term agreements at a much lower cost known as 'base' load.

These short term purchases of electricity at a high price are invoked as justification for making large number of hydropower projects. The Chief Minister of one state told this author that he needed to build more hydropower projects because he had to buy power at rates as high as Rs. 8.50 per unit. In taking such decision it is ignored that long term prices are much lower. It is necessary to make an estimate of the long term demand of power (Please see *Chapter 10: Economic Value of Hydropower* and *Chapter 11: Critique of FERC's Approval to Klamath Hydroelectric Dam*).

Value of Social Benefits

Our planners do not make any assessment of 'benefits' secured by the economy from increased generation of electricity by a particular project. It is assumed that benefits are huge. This approach can lead to making of projects that are economically harmful in the long run.

Two problems are involved here. One, our planners do not assess the social benefit from a particular project. Two, the demand is overstated.

The social benefit from electricity is the difference between the cost of production from (1) the proposed project and (2) the next best alternative source. Let us say a project is designed to produce 1 million units of power at Rs 2 per unit. In absence of the proposed project, the same 1 million units would have to be produced by alternative sources, say at Rs 3 per unit. The benefit from the project is thus estimated to be Rs 1 million (at Rs 1 per unit). Our planners do not make such a calculation at all. Therefore, we do not know whether the benefits from a project are justified by the costs.

Secondly, the demand for power is overstated. Electricity is sold to the consumers at a price lower than the cost of production. Farmers are provided with free power is some states. The demand at the cost of production will be less hence the need to produce power will also be less (Please see *Chapter 9:*

Cost-Benefit Analysis and *Chapter 10: Economic Value of Hydropower*).

Value of Electricity during Monsoons

Run-of-river hydropower projects generate most electricity when the demand is less—during the monsoons—due to low agricultural demand. In the winter and summer, when energy demand is higher, river flows are at their lowest and they produce less power. This is intra-season variation in demand for power.

On the other hand, storage-based hydropower dams can produce more electricity during peak hours when the demand is high for lighting etc. Water is stored during non-peak hours and used to produce power during peak hours. This is intra-day variation in demand for power. This benefit of hydropower is repeatedly mentioned in Government documents such as in the Integrated Energy Policy Report of the Planning Commission.

Thermal electricity plants cannot 'switch-on' and 'switch-off' with ease. They have to continue producing electricity once the boilers are fired. They produce large amount of non-peak electricity along with peak electricity.

Unfortunately, the low value of benefits from electricity generated during monsoons is ignored in this assessment while the high value of benefits during peak hours is recognized. Thus we adopt lopsided valuation of electricity (Please see *Chapter 3: Economic Analysis of Removing Four Dams on the Lower Snake River* and *Chapter 11: Critique of FERC's Approval to Klamath Hydroelectric Dam*).

Comprehensive assessment of costs and benefits

Our planners aim to produce electricity to secure people's welfare—to provide youth with lighting to study and to the homemaker to run her kitchen appliances. But this is only one aspect. 'Welfare' is truly a holistic concept. It includes employment and income as well. Electricity may not add to welfare if people are deprived of income in the process of its generation.

For example, a hydropower project may displace 100 households. These households may loose income from their agricultural lands. It is possible that the loss of welfare due to deprivation of this income may be more than the gain of welfare from consumption of electricity—at least as far as these affected persons are concerned. Projects must, therefore, make a holistic assessment of all benefits and costs from generation of electricity. In the study of Lower Snake River dams, it is shown that removing the dams and replacing their power with clean energy (energy efficiency and wind power) would not impede economic growth, and could create as many as 15,000 long term jobs. The economic assessment of hydropower projects made by us does not assess these impacts holistically. For example, the employment generated in the projects is mentioned but the employment lost in submergence of agricultural lands is not mentioned (Please see *Chapter 3: Economic Analysis of Removing Four Dams on the Lower Snake River*).

There is a huge political determination to harness hydropower in India. Hill states like Arunachal Pradesh, Himachal, Sikkim and Uttarakhand find generation of hydropower a profitable venture. Bhutan is very comfortable selling surplus power to India. The dams recently removed in the United States had similar benefits. There is a huge effort to remove the Klamath hydropower dam. The Edwards dam was producing hydropower when it was decommissioned. We should not be under the impression that only obsolete dams are being decommissioned.

Demand Side Management

The objective of producing electricity is to secure people's welfare. There are many ways of securing this. It is possible that the benefits to the society may be more from Demand Side Management than from production of more electricity. The Oregon Public Utility Commission rejected PacifiCorp's request for establishing a power plant saying that PacifiCorp had overstated its need for power and should pursue short-term market purchases, renewable energy, demand-side management and efficiency measures instead. Such assessment of alternatives is not made by our authorities (Please see *Chapter 11: Critique of FERC's Approval to Klamath Hydroelectric Dam*).

Storing water without disturbing free-flow of rivers

An important function of large storage dams like Bhakra, Rihand, Nagarjuna Sagar, Sardar Sarovar and Tehri is to store water during monsoons and use it for drinking or irrigation during lean periods of winters and summers. A dam is made on the main stream of river flow. The river bed is converted into a reservoir. The free flow of river is wholly destroyed in this stretch. This is 'on-line' storage of water involving holding back water within the river channel by a dam.

There is a better alternative. Water of the river can be diverted 'off-stream' for storage away from the river flow. The environmental impact of such off-stream projects is less severe because the functions of the river are left unaffected—only the quantity of water is reduced during the removal periods. The main river can flow uninterrupted.

Such a side-stream skimmer has been built by the City of Fredericksburg in Virginia. The Embrey Dam was providing drinking water to the city of Fredericksburg through a canal. It blocked the flow of Rappahannock River and diverted water into a canal. This dam has since been removed. Water is now being provided to the city by the Motts Run pumping facility. This facility pumps out water from the river during high flows without obstructing the main flow. Water from River Rappahannock is pumped into the Motts Run Reservoir during high-flows and then used for supplying water to the city during lean periods. River Rappahannock flows uninterrupted. This model can help us remove large storage dams like Bhakra, Sardar Sarovar and Tehri and restore free flow of rivers while maintaining supply of water during lean periods. We must study the possibility of making off-stream artificial reservoirs and removing these large dams (Please see *Chapter 6: Removal of Embrey Dam* and *Chapter 8: The Motts Run Water Treatment Plant*).

Fig. IV : Pumping station on Rappahannock River for diverting water to Motts Run artificial reservoir.

Off-stream projects are, of course, more expensive. They require building of inlet and outlet in addition to storage reservoir. Suitable sites may not be easily available for making such storage. Further, the high flows that are diverted also have important ecological functions. They spread water over large areas and help in ground water recharge and flood-recession agriculture. These costs have to be internalized. Thus, off-stream storage is not a panacea. It is only an alternative that must be examined dispassionately (Please see *Chapter 12: Impacts of Hydro dams on environment*).

Benefits from fisheries and other 'non-power' uses of river waters

The objective of producing hydropower is to secure welfare of the people by providing electricity. The same welfare is also

Fig. V: The Motts Run artificial reservoir.

secured by recreation or other non-power benefits such as grazing, harvesting of sand, protection of endangered species etc. It is possible to place money value on many of these benefits. Once this is done, it is possible to assess the benefits from electricity and compare the same with the benefits from recreation etc. Unfortunately, our planners assess the benefits from hydropower but ignore the benefits from recreation etc. It is possible that the loss from recreation etc. may be more than the gain from generation of electricity. In such case we are implementing projects which have a negative impact on the people. The impact of a proposed project on welfare of the people must be made in a straight forward way. It cannot be assumed that benefits from generation of electricity are greater from the loss of recreation etc. (Please see *Chapter 17: Non-use value of free flowing rivers*).

A large number of dams are being removed in the United States because the FERC requires dam owners to install fish elevators. It was thus assessed that removing the Glines and

Fig. VI: Fish elevator at Conowingo Dam.

Elhwa dams would be better than installation of fish ladders (Please see *Chapter 4: Non-use Values lead to the Removal of Elwha Dam*).

There is a need to comprehensively assess the benefits from recreation etc. Increase in recreation from boating in hydropower reservoirs, for example, is offset by loss of recreation from rafting and kayaking (Please see *Chapter 3: Economic Analysis of Removing Four Dams on the Lower Snake River* and *Chapter 12: Impacts of Hydro dams on environment*).

The importance of examining alternatives ways of securing welfare is brought home by the denial to Wilkes-Barre Inflatable Structure. This inflatable dam was proposed to be built on the Susquehanna River for flood control and enhancement of recreational facilities. The authorities held that the objective of enhancing recreational facilities can be accomplished through a number of non-structural alternatives which do not obstruct flow of the river. They found that recreation could be secured without building the structure. They denied permission to build the dam. The order determined that the economic value from the proposed dam is less than the economic value that can be created from riverfront development. This is the true meaning of 'alternatives' that are required to be considered as a part of Environment Impact Assessments under the EPA in India, but rarely done in practice (Please see *Chapter 7: Denial for Wilkes-Barre Inflatable Structure*).

Absolute values of free-flow of rivers

People in the West are moving to preserve the free-flow of their rivers for maintaining certain absolute values like that of the National Flag. One does not tear up the flag to make garments even if shivering in cold. Free-flow of rivers contains four types of such absolute values.

River as a living deity

During the removal of the Embrey Dam in Virginia, Senator John Warner quoted a passage from Ezekiel in the Bible "There will be a very great multitude of fish, because there waters go there; for they will be healed and everything will live wherever the river goes." He drew sanction from dam removal from religion. Hindu scriptures lay a much greater stress on the salvation-giving powers of the rivers—Ganga in particular. We place a kalash full of water near the worship as representative of the rivers. Unfortunately, we appear to negate these religious values and are single-mindedly focused on the economic exploitation of the river waters (Please see *Chapter 6: Removal of Embrey Dam*).

A review of environmental impacts of dams in New Zealand speaks of 'self-essence' of a river. It points out that reductions in flow will influence the 'sound, smell and overall mood' of the river. This is close to valuing the river as a living deity as in our culture. These values are degraded by making of dams. This self-essence is same as the veneration of rivers as Mother in our tradition (Please see *Chapter 12: Impacts of Hydro dams on environment*).

Article 25 of our Constitution provides that "all persons are equally entitled to freedom of conscience and the right freely to profess, practice and propagate religion." This right is an absolute right that cannot be overridden by economic benefits. A similar situation prevails in the United States with respect to the cultural rights of Tribes. A report relating to the Elhwa Dam says: "the benefits from fishing do not capture the enormous benefits to the Tribe, who place tremendous cultural and spiritual importance on the river, its fish, and the historic sites inundated by the dams' reservoirs." The absolute right of the Tribes superceded any economic benefits that may accrue from retaining the dam. This observation is relevant for us. All our rivers are worshipped as deities. This worshipful attitude will certainly be hit by diverting the river into tunnels and canals. This enormous loss is almost never examined in our EIAs. This is especially relevant to the dams being built on the Ganga, which has been declared to be our National River. Rivers are an important aspect of the spiritual heritage of India. This must be incorporated in the decision making process (Please see *Chapter 4: Non-use Values lead to the Removal of Elhwa Dam*).

Endangered species

The Endangered Species Act of the U.S. requires the Government to protect endangered species. This mandate overrules economic considerations. Our Constitution similarly requires us to protect wildlife. We have made Tiger Reserves for this purpose. We forego economic benefits from forests in order to preserve Tigers. We do this without undertaking a Cost-Benefit Analysis. That is, we preserve the tiger irrespective of the economic cost of doing so. Our rivers stand on a much higher footing than the tiger. We worship our rivers as mother or as a deity. Therefore, we should apply the same logic and forego the benefits from hydropower and irrigation in order to preserve our rivers.

These non-monetary or absolute benefits should be weighed in parallel with benefits that lend themselves to monetization. We appear to be making two mistakes in this regard. One, we do not explicitly state the non-monetary benefits like impact on worshiping. Two, we do not make correct assessment of monetary benefits.

Wild and Scenic Rivers

The Wild and Scenic Act (WSA) was enacted in the United States to preserve free-flowing waters in specified rivers or part thereof. The idea is that certain parts of the rivers are of incomparable aesthetic value and must be preserved irrespective of the economic costs that are involved.

"Free-flowing" is defined in WSA as "existing or flowing in natural condition without impoundment, diversion, straightening, rip-rapping, or other modification of the waterway." This definition is important as it precludes building of Run-of-River dams that impound, divert and modify the flow of our rivers. Our MOEF, on the other hand, routinely assumes that making barrages for hydropower dams does not affect the flow of the river (Please see *Chapter 15: The Wild and Scenic Act*).

The WSA enables the Federal Government to overrule the rights to exploit the river for irrigation or hydropower vested under State Law. Our Environment Protection Act and Forest Conservation Act provide similar overriding powers to the Union Government. These powers can be used to protect our rivers from over-zealous exploitation by State Governments (Please see *Chapter 15: The Wild and Scenic Act*).

The activities disallowed under WSA include, among others, the following:

- Making a hydroelectric project on a tributary of a river.
- Blocking a tributary of a river for irrigation purposes.

- Construction of a boat landing (Please see *Chapter 16: The Wild and Scenic Act in Operation*).

The essence of WSA is to preserve the river in her natural condition.

Non-use value of free-flow of rivers

The argument for protection of free-flow of our rivers made in the preceding paragraphs is based on certain absolute values. The second level of argument is that proper economic valuation of the free-flow of rivers should be made. Hydropower projects are made to promote welfare of the people. The production of electricity would enable the homemaker to light up the temple in the house. This is 'use' value of river waters. We use the electricity that is produced. The satisfaction one obtains from taking a bath in the river or from undertaking a whitewater rafting expedition are likewise use values.

There are other 'non-use' values. Many taxpayers in the country support the protection of tiger even though they may never visit the reserve to see the tiger. They derive certain satisfaction from the knowledge that the tiger is not extinct and that it survives. This is 'non-use' value. A similar non-use value is obtained from the knowledge that free flow of a river is maintained. Just as a taxpayer is happy that the tiger is not extinct, similarly he is happy that free-flow of river is maintained.

The river has both 'use' and 'non-use' values. Making a hydropower dam provides certain use values. Free-flow of river also provides certain use values such as of bathing and also non-use values. Both these values can be reduced to money terms and can be assessed.

Economists in the United States have calculated these non-use values. They have found that non-use values of free-flow of river waters are much larger than use values. They are planning to remove certain dams based on such calculation. The Elwha Dam is slated for removal because it was found that a large non-use value would be obtained from removal of the dam (Please see *Chapter 4: Non-use Values lead to the Removal of Elwha Dam*).

The non-use value of rivers is probably many times more in our culture that in the American culture. We worship rivers as mother or as a deity. The loss of this non-use value should be calculated when the decision to make a dam is made. Calculation of non-use values of Ganga and Narmada Rivers can similarly lead to removal of Tehri and Sardar Sarovar dams.

Reservoirs also provide use values, especially to the tourists. On the other hand free-flowing rivers provide use values to pilgrims who take bath in the rivers. Both should be dispassionately assessed. But in practice, the use value obtained by pilgrims from free-flow of rivers is ignored.

Depletion Leads to Higher Non-use Value

The fact that a species is endangered means that economic value cannot be created from its 'use'. If the number of tigers is small, it will not be possible for a large number of people to derive use value from watching them. At the same time, depletion leads to increase in non-use value. Let us say, tigers were roaming the countryside aplenty. People would hardly derive any non-use value from them just as one hardly derives any non-use value from watching the goat. The non-use value increases as the number of tigers declines and the species moves towards extinction. If there were, say, only 10 tigers in the world, the non-use value of each tiger would be astronomical—akin to the value of a Van Gough painting.

This concept applies to our rivers. The non-use value of free flow of over rivers will increase as more dams are built. This benefit can be calculated just as one can calculate the benefit from consumption of electricity. This mistake in accounting leads to a decision of building dams for generation of electricity (Please see *Chapter 17: Non-use value of free flowing rivers*).

Non-quantitative Values

We have discussed the absolute values and non-use values above. A third type of value is 'non-quantitative'. It is not possible to place a value on every item. For example, how much value would a child place on the mother's smile? Our rivers provide many such values that defy quantification. These include providing habitat to endangered species, providing lingwas for the dead souls, providing salvation from rebirth and biodiversity. On the other hand, dams impose many non-quantifiable costs upon the people. These include trauma due to earthquakes and landslides, loss of soul due to displacement or due to submergence of a temple and trauma due to increased incidence of water-borne diseases such as malaria.

The Office of Management and Budget of the U.S. Government has issued guidelines for making Cost-Benefit Analysis. These guidelines state that a schedule of non-quantitative costs and benefits should be made and provided to the policy makers along with a Cost-Benefit Analysis. This schedule may be in form of a list. The EIA can provide such a list of impacts. Unfortunately, our projects ignore the non-quantitative costs of dams and benefits of free-flowing rivers (Please see *Chapter 9: Cost-Benefit Analysis*).

Assessment of Alternatives

It is customary for our regulatory agencies to provide a single Cost-Benefit Analysis. It is considered adequate to show that benefits from a project are greater than the costs. The guidelines issued by the Office of Management and Budget state that such one-scenario analysis is inadequate. It is necessary to examine various alternatives. For example, it is inadequate to show that benefits from a particular hydropower dam exceed the costs. It is necessary to compare these benefits and costs with other alternatives such as demand side management, solar power, micro hydel etc. so that second-best projects are not implemented. A particular hydropower project may have Cost-Benefit Ratio of, say, 2.0. It should not be implemented on this basis alone. It is possible that Cost-Benefit Ratio of demand side management may be 2.5. In that case, demand side management must be implemented instead of making a hydropower project (Please see *Chapter 9: Cost-Benefit Analysis*).

Valuation of Services Provided by River Waters

The reader would be familiar with the terms 'capital' and 'revenue' expenditures. Capital expenditures are those which provide services for an extended period such as planting of an apple tree. Revenue expenditures, on the other hand, are those that provide services for one event such a purchase of a box of apples. A similar distinction applies to natural resources. A forest provides carbon sequestration for its lifetime while it provides timber as a single event. Capital expenditures are generally considered 'good' while revenue expenditures are considered, relatively speaking, 'bad'.

This distinction applies to our rivers. Rivers provide certain long term services such as sediment renourishment to our coasts, biodiversity, aesthetic value, etc. They also provide 'short term' benefits like hydropower. Following the above distinction, we should give priority to long term environmental services instead of the short term benefits from hydropower and irrigation. Such distinction should be built into our Cost-Benefit assessments. *Chapter 18: Economic Value of New Jersey Rivers* makes an assessment of the capital value of forests of New Jersey. The study shows that freshwater wetlands have a value of $11,802/acre/year and riparian buffer has a value of $ 3,500 which is much higher than forest land at $1,714. This can serve as a model for valuing free-flow of our rivers.

Sediments

An important function of rivers is to transport sediments to the coasts. Making of dams impedes this process. Sediments are trapped in the reservoirs. They are flushed at regular intervals in smaller barrages but this intermittent flushing yet changes the flow dynamics of sediments.

Rise in Tail End Level

Trapping of sediments in the reservoirs leads to raising the reservoir water level by a few inches every year. This sediment build-up has elevated the water level behind a levee in Lewiston, Idaho. As of 2002, the 5 feet of levee protection had been reduced to only 2 feet. High river flows would put Lewiston at risk of flooding. Similarly, the bed level upstream of the Waihopai dam in New Zealand has increased by 25 meters since 1926 (Please see *Chapter 3: Economic Analysis of Removing Four Dams on the Lower Snake River* and *Chapter 12: Impacts of Hydro dams on environment*).

The large numbers of embankments made by our Government for flood control lead to deposition of sediments in the river bed and raise the level of the same. The devastating impact of this was seen in the Kosi floods in Bihar in 2008. This writer has seen many EIAs of reservoir-based hydropower projects. The cost due to raise of the level of the tail end reservoir is never mentioned to the knowledge of this writer.

Reduced Capacity of Storage

Trapping of sediments fills up the reservoirs and reduces the live storage capacity. The Condit Dam on the White Salmon River of Washington State, for example, is full with sediments which has reduced the production of power. This cost must be internalized in our economic assessments.

Flushing of Sediments is not Enough

Sediments are flushed from Run-of-River (ROR) hydropower dams periodically. Indian experts claim that the total flow of sediment in the cycle of, say, 15 days, is not affected hence there will be no impact on the ecological function of sediments. *Chapter 12: Impacts of Hydro dams on environment* of this volume rebuts this view. It points out that erosion of river banks takes place when the sediment level is low and deposition takes place when it is high. Thus the downstream areas are, at the least, subjected to repeated bouts of deposition and erosion and are made unstable. Deprivation of sediments also threatens the stability of downstream bridges and other structures.

Cost of Removal of Sediments

Dams have a finite life of about 50-70 years. They require extensive repairs after this period. It has been found in the U.S. that many times the cost of repairs becomes exorbitant and makes it more profitable for dam owners to remove the dams. But removal faces a hurdle. Large amounts of sediments are trapped in the reservoirs. They are released when the dam is removed and can cause huge ecological destruction in downstream areas. In order to prevent such costs, the possibility of dredging and transporting the sediment through pipelines to the sea was considered during removal of the Elwha Dam. The silt, gravel, clay and water deposited behind the Embrey dam was proposed to be propelled through a pipeline to a 13-acre dredge containment area. We will face similar costs when we remove our dams. These costs must be built into the economic assessments (Please see *Chapter 4: Non-use Values lead to the Removal of Elwha Dam* and *Chapter 6: Removal of Embrey Dam*).

Environmental impacts of removal of dams are now known. Say a particular impact, such as discharge of sediments, takes place at the time of removal. This impact occurs in reverse during making of a dam. If release of trapped sediments has an environmental impact, the trapping of the sediment will also have an impact. Dam removals have released toxins. These toxic substances get accumulated behind dams and their increased concentrations can lead to extensive damage when released (Please see *Chapter 1: Problems of Dam Removal*).

Coastal Erosion

Another impact of trapping of sediments in reservoirs is that of coastal erosion. Indian rivers carry large amounts of sediments to the ocean. These sediments renourish the coasts and help them counteract the erosion that takes place naturally due to the cutting action of the waves. Trapping of sediments prevents this renourishment and subjects the coasts to accelerated erosion. This can be seen happening at Ganga Sagar Island in West Bengal.

Fig. VII : Coastal erosion in Tunisia contributed by trapping of sediments of Medjerda River.

Fig. VIII : Coastal erosion in Tunisia contributed by trapping of sediments of Medjerda River.

Fig. IX : The Medjerda River in Tunisia Converted into a Canal.

This problem is not very prominent in the discourse on removals in the United States. However, Prof Jim Puzzuto of Department of Geological Sciences, University of Delaware told this writer that building of dams on the Mississippi River has contributed to erosion of coastal lands. The delta is sinking and this was one of the reasons of the devastating impact of the Katrina Hurricane on the city of New Orleans.

This writer had an opportunity to visit Tunisia in Africa. This country has build an intricate network of dams, canals and pumping stations on its main Medjerda River. As a result the sediments are wholly trapped in the reservoirs and coastal erosion has accelerated.

Water Quality

Change in Timing of Discharge

The demand for electricity is high during mornings and evenings which are known as 'peak' hours lasting about 3 hours each. An important benefit of hydropower is that its generation can be started and stopped with ease in order to produce electricity during peak hours. Our Central Electricity Authority approves ROR dams 'with pondage' and allows them to modify intra-day discharge of water to meet peak load demands. Proponents of hydropower contend that this intra-day variation does not affect the basic characteristic of the flow of river. The total amount of water flowing during a day remains unchanged. There is only a small change in the timing of the discharge. Question arises whether such change in timing of discharge affects the flow of water or not?

This question was raised before the U.S. Appeals Court. A hydropower plant was already in operation. The producer sought to change the configuration of turbines. This would admittedly lead to change in timing of discharge of water. The request was denied. The Court held that a "change in discharge" takes place if the timings are changed even though the "total daily volume" remains unchanged. The Court asked, why 'discharge' should be measured on a daily, as opposed to on a weekly or even yearly, basis? If variation between seasons is considered a change in discharge, then variation between hours of a day also amounts to change in discharge. There is no reason to ignore change in discharge when it occurs on a daily basis while accounting for change in discharge on a seasonal basis. It was held that change in flow takes place due to intra-day variations (Please see *Chapter 13: Impact of Hydro Turbines on Water Quality*).

The 'ROR with Pondage' hydropower dams cause change in the flow of water between peak- and non-peak hours. This has a decisive impact on aquatic life. Fish and invertebrates may find an area suitable during peak hours but not during non-peak hours. They may not be able to adjust to this change.

Impact of Cascade of Dams

River water rubs against stones and air, and interacts with earth and flora and fauna during flow. This interaction is prevented by diversion of flow through tunnels, canals and reservoirs in hydropower dams. However, the water regenerates these same qualities as it flows downstream after discharge from the hydropower dam. Such regeneration will not take place when we build a cascade of dams and no free flow of water is obtained downstream (Please see *Chapter 12: Impacts of Hydro dams on environment*).

Aerators for Maintaining Water Quality

This writer had an opportunity of visiting the Conowingo Dam in the United States. The officers informed that the Federal and Maryland State authorities required the dam owners to maintain high level of Dissolved Oxygen in the discharged waters. The dam owners installed aerators in the turbines to secure this objective. Air was pumped into the water through

Fig. X : Air Suction Inlets for Aerators at Conowingo Dam.

blade of the turbines in order to increase the DO. The cost of installing this equipment was about $4 million per unit. It also led to one to two percent reduction in generation of electricity because wind shear reduces the speed of the rotor.

Health Hazard of Methane Bubbles

The pressure inside turbines forces gasses like methane into water in form of bubbles. These bubbles can get into various parts of the body. Major fish kills have resulted from such super-saturation of gas. Consumption of downstream water can have similar effect on human beings (Please see *Chapter 12: Impacts of Hydro dams on environment*).

Savings of Water Treatment Costs

Free flow of river improves the water quality of the river and reduces the cost of water treatment facilities. Deprivation of free flow by hydropower projects, therefore, imposes a higher cost of treatment. On the other hand, trapping of sediments can reduce the cost of treatment. This economic cost and benefit must be internalized in projects (Please see *Chapter 3: Economic Analysis of Removing Four Dams on the Lower Snake River*).

Questions of Equity

Dams have varying impacts on people. They benefit one group while they impose costs on another group. This different impact has a social cost.

Say cost of production of 1000 KwH electricity is Rs 1,000. This electricity is sold for Rs 2,000 hence the project is profitable. On the face of it, such a project would be socially beneficial. The society will make a net gain of Rs 1,000 from making of the project.

But it is possible that the cost of Rs 1,000 is borne by a poor household whose land is submerged in the hydropower reservoir. On the other hand, the benefit of Rs 2,000 is secured by the rich man for running his air-conditioner. In such a situation the total happiness or well-being of the society may be reduced. The pain borne by a poor man from loss of Rs 1,000 may be much greater than the gain of Rs 2,000 obtained by a rich man. It is necessary, therefore, to look at who gains and who looses.

The study of removal of four lower Snake River dams points out that the benefits of removal will accrue to hikers, birders, boaters, campers, and other visitors while the losses will accrue to urban users of power. In other words, the making or breaking of a dam hides a tension between nature-lovers and urbanites. Making of hydropower dams on our sacred rivers like Ganga hides a similar tension between pilgrims—who stand to loose—and urbanites—who stand to gain. The Tehri Dam, for example, imposes costs on pilgrims at Haridwar and poor people living around the reservoir rim while it provides drinking water to Delhi. This must be explicitly mentioned. It is not enough to state that total benefits are greater than total costs (Please see *Chapter 3: Economic Analysis of Removing Four Dams on the Lower Snake River*). The Office of Management and Budget of the U.S. Government explicitly requires assessing authorities to specify if gains and losses occur to different groups.

Another aspect of equity is inter-generational. Dams have finite life even though some may last for 100 years or more. At some point future generations will inherit non-functioning dams along with broken rivers. The present generations will benefit at the cost of these future generations. This tradeoff should be explicitly stated (Please see *Chapter 1: Problems of Dam Removal* and *Chapter 9: Cost-Benefit Analysis*).

Withdrawal of energy from water deprives another living being

Hydropower is presented in India as a win-win proposition for all players. Claim is that water is not consumed while energy, that was previously being wasted, is harvested for productive purposes. The key assumption here is that downward flow of water in a river is a 'waste' of energy.

Recent research in the area of 'River Continuum' points otherwise. It is pointed out that the living organisms develop to maximize the utilization of energy from the free flow of rivers. Flora and fauna adapt in a way to maximize energy consumption. When, therefore, we remove energy from water through hydropower dams, we necessarily deprive other living beings of that energy.

Flow of the river involves dissipation of kinetic energy. This rate and nature of dissipation is affected by width, depth, velocity, flow volume and temperature. Hydropower projects necessarily involve change in these physical characteristics. Thus living organisms have to readjust to the new circumstances.

A free-flowing river has a great variation of ecological conditions. This variation is the trigger for the creation of biodiversity. Variation of conditions leads to generation of more diverse species to utilize these diverse forms of energies. Thus making the river flow 'stable'—such as by converting free flow into a reservoir—stabilizes the system and leads to low biological diversity (Please see *Chapter 14: The River Continuum Concept*).

EIAs done in India usually list the endangered species of flora and fauna. If none are present then it is assumed that a project will not have any negative impact on environment. A similar situation prevailed in the Wilkes-Barre proposal. No Federally listed or threatened or endangered species were known to occur within the project area. But the U.S. Fish and Wildlife Service did not stop here. It pointed out many other impacts such as obstruction to fish migration, deterioration of water quality, alteration of riparian wetlands and shallow water habitats and sediment deposition and erosion. Our State Forest and Wildlife Departments should likewise go beyond endangered species and look at the impact on wildlife more holistically (Please see *Chapter 7: Denial for Wilkes-Barre Inflatable Structure*).

Economic value of river banks

We are building a large number of embankments along our rivers as a flood control measure. These embankments can also double up as a highway as proposed in the Ganga Expressway project along the Ganga in Uttar Pradesh. The assumption is that naked riverbanks do not provide any economic benefits. They are 'unproductive'.

The paper *Economic Value of River Banks* at Chapter 19 in this volume disproves this view. It shows that riverbanks perform a wonderful function of cleaning the river waters. River banks are efficient at processing organic matter and sediments. Finer sediments are removed by reactions with soil. Bacteria and fungi living in stream bottoms consume nutrients brought by polluting discharges and convert them into beneficial compounds. Plants in river banks absorb harmful chemicals. They also provide aesthetic, biodiversity and carbon sequestration services. The cost of cleaning water through riverbanks is reckoned to be less than that from water treatment plants. Authorities calculate values of property saved etc. by embankments but do not calculated value of these services provided by river banks.

Embankments on low-gradient floodplains have large impact on aquatic life because they deprive fish from access to large flood areas (Please see *Chapter 12: Impacts of Hydro dams on environment*).

Humble submission

The main point is that the purpose of economy is 'welfare', not economic growth *per se*. We should stay focused on welfare. These include non-use values, non-quantitative values and equity. Internalizing these impacts of hydropower projects would often tip the scales against the projects.

PART-I

WHY DAMS ARE BEING REMOVED IN THE UNITED STATES

1

Problems of Dam Removal[1]

Jock Conyngham, J. Craig Fischenich and Kathleen D. White

Editor's Introduction: The U.S. Army Corps of Engineers was previously involved in making dams. Nowadays it is increasingly involved in removal of dams. This paper by officers of the Corps' Engineer Research and Development Center explains the problems of removal. This paper is relevant for India in two ways. One, the dams being made by us now will incur the costs mentioned here in future. Our present generation is imposing these costs on our future generations. Two, the impacts mentioned here work in reverse in dam building. We should recognize these costs now if the U.S. is recognizing these costs at the time of removal. The following points may be noted from this paper:

---

1. Adapted from "Engineering and Ecological Aspects of Dam Removal—An Overview" EMRRP, September 2006.

- Of large dams, 85 percent will have exceeded their design lifespans by 2020 or soon thereafter. Lifespan of dams being built by us will likewise be about 50 years. Many dams have filled with sediment long before lifespan assessments. For example the Milltown Dam on the Clark Fork River in Montana got filled with mine waste in a large flood in 1908, the year after it was completed. Hydropower dams are not 'permanent' as made out to be. The costs at the time of removal should be internalized.
- Dam removal often costs far less than the restoration of deteriorating structures. The deterioration of dam structures is a significant cost *in the long run.* Dam companies account for maintenance costs for 30-year life on which their economic viability is worked out. The costs to be incurred after this period are left unaccounted. The lease agreements with dam companies that this author has seen do not provide for dam removal costs. Implicitly these costs are to be borne by the society at a later date while the benefits are reaped by the dam companies at present.
- Dam removals have caused occurrences of released toxins or nutrients, channel instability, downstream sediment impacts and invasive populations.
- Toxic substances will flow to the ocean in small concentrations along with flow of river water in an free-flow system. The same toxic substances get accumulated behind dams and their increased concentrations can lead to extensive damage when released. In such cases, active sediment removal is needed.
- Channels are deinstabilized twice by dams. Once when the dam is built and banks are inundated; and again when the dam is removed and banks are let free.
- Invasive populations are dealt with gradually by nature. The upstream areas develop resistance to such invasions.

Dams removals allow a sudden invasion of the same species and can become devastating because the organism that resist such intrusion have not had the time to develop.

- There have been a large number of dam failures in the United States. There is always some risk of dam failure. The economic assessments do not internalize this cost. It is assumed that probability of dam failure is zero. This is clearly not the case.
- A study for Wisconsin found that that cost of removal was about one third the price of dam repair.
- Higher value given to nature leads to dam removal. The Indian ethos of *Vasdev Kutumbakam* leads to a similar conclusion.

Overview

Decommissioning and removing dams has emerged as one of the central foci of the new millennium for infrastructure management, river conservation, and the restoration of fisheries populations. Anadromous, catadromous and adfluvial species are especially impacted by dam decommissioning and removal. It represents arguably the most powerful tool and largest opportunity for restoration of aquatic ecosystems and communities that currently exists. Several phenomena underlie this development:

- *High dam densities and the aging of dam infrastructure.* Of large dams, 85 percent will have exceeded their design lifespans by 2020 or soon thereafter (Federal Emergency Management Agency 2001). Though inventories are poor, dams exist at much higher densities than many realize.
- *Threats or occurrences of dam failures.* In 2000 and 2001, 520 dam incidents and 61 dam failures occurred; the American Society of Civil Engineers gave dam management and safety a grade of "D" in the last two editions of its "Report Card for America's Infrastructure." Although inventories are incomplete (with 11 states having

no inventory at all), 2,100 US dams are categorized as unsafe and almost 10,000 as high hazard potential, and both categories show significant growth in recent years (ASCE 2002).

- *Failure of traditional restoration*. The mixed success or outright failure of expensive efforts to protect and recover various threatened and endangered species as well as critical prey populations, e.g. the herrings, has received much attention in recent years. The effect of dams on both upstream and downstream migration success is usually cited as a central factor.
- *Improved knowledge of aquatic ecosystems and processes.* Recent advances in knowledge and/or awareness of geomorphology and the ecology of regulated rivers have increased attention on the effects dams have on sediment budgets, hydrology, water chemistry, and life history needs.
- *Economics.* Dam removal often costs far less than the restoration of deteriorating or sub-standard structures. One review of 30 case studies in Wisconsin indicated that removal cost on average about one third the price of dam repair (Born et al. 1998). In addition, many of the values that dams have historically provided, such as hydropower, have been superceded by the direct or opportunity costs of values now considered important (liability, recreational values, large populations of important recreational or commercial fish, etc.).

Despite current enthusiasm, the process of dam decommissioning and removal cannot be viewed as a simple environmental panacea but, rather, as a set of analyses, decisions, and trade-offs with both beneficial and adverse impacts across the spectrum of ecological, social, and economic concerns. Dam removal can require considerable technical expertise and carries potential risk of physical instability, ecological or economic impacts, and local backlash. Recent dam removals have caused occasional but significant occurrences of released toxins or nutrients, channel instability, downstream sediment impacts, invasive plant infestations, and increased risk of ice damming, often despite demanding regulatory overview. Scientists and regulators have expressed concern about current removal practices and requested technical guidance to delineate determination of dam fate, the suite of relevant issues, and the appropriate selection and sequencing of tools for dam removal and associated restoration where indicated. Due to the poor condition of many dams as well as restoration mandates and goals, guidance on efficient and cost-effective project implementation stands out as a critical further need.

Planning Considerations

More than 68,000 large dams and an undetermined total number of dams, probably exceeding 2,500,000 structures, exist nationwide (NRC 1992). With a large percentage of these structures, particularly the smaller ones, in undetermined but deteriorating shape and nearing the end of or having passed their design lifespan, the United States faces a pressing public policy need. Affected populations of threatened and endangered species as well as prey populations that help to enrich and drive entire freshwater and coastal ecosystems (such as various herring and shad species) add significant economic and ecological stakes to this problem.

Dam removals to date, however, have often been targets of opportunity or crisis. Born et al. (1998) suggest that public safety and the desire to save costs of repair usually drive removals; only rarely is watershed or fisheries restoration a catalyst for dam removal. A number of projects, however well-intentioned, have resulted in adverse impacts and sometimes public reaction. Griffin (2001) describes one case of local backlash at a proposed removal in Maine that ultimately catalyzed unsuccessful attempts to pass legislation to slow removal efforts. Many removals to date have cost more money or taken more time than necessary, and the full costs are often unknown. Clearly, a proactive, planning-driven, and technically robust approach to a mounting problem is indicated.

Benefits and Costs of Dams

Dams have provided and continue to provide a diverse suite of services and values to owners and society. These include:

- Impoundment-based recreation
- Farm pond and firefighting water
- Flood control
- Water supply for irrigation, residential, and industrial uses
- Hydropower
- Protection from ice damming downstream of structure
- Management of tailings, toxic sediments, or excessive nutrient loading
- Historic and archaeological values of structure and/or associated buildings

As described earlier, dam removal advocates have on occasion catalyzed local backlashes, particularly in cases when economic interests came into play or where the process was relatively well-advanced when local stakeholders learned of the impending work. Effective analytic protocols, alternatives analyses, lucid metrics for decision-making, and clean lines of communication are required for efficient project development and implementation.

Dam costs occur at various scales and must include consideration for:

- Risk of failure (Figure 1.1).
- Ecological damage from alteration of hydrologic, thermal, and chemical regimes.
- Ecological damage from significant alteration, in most cases, of sediment regimes, usually resulting in upstream aggradation and downstream incision (Collier et al. 1996). Dams that regulate the hydrologic regime or divert flows, however, can result in downstream aggradation because

Figure 1.1 Rockfish Creek Dam failure, NC, 2003

of the system's inability to route sediment supply from tributaries.

- Ecological damage to riparian zones from decoupling of the channel and floodplain by flow-regulating dams (Magilligan et al. 2003).
- Interruption of requisite upstream and downstream movements of populations, altered predation regimes, habitat fragmentation, and increased risk of exotic organisms due to mixing of lentic and lotic habitats. For instance, unobstructed stream and river reaches have been reduced 91 percent by large dams in the North Atlantic regions (Lary and Busch 1997).
- Chronic and high liability (related to both risk of failure and injuries or deaths due to the structure) and maintenance costs.

Dam removal itself has ecological and economic costs and impacts. These costs must be considered relative to the benefits of removal in alternatives assessments (Bednarek 2001; Schuman 1995, 2002). Examples of potential impacts include:

- Release of excessive sediments. A recent analysis of five removals in Wisconsin showed sediment loading from 170 to 1120 percent of normal sediment budgets (River Alliance of Wisconsin/University of Wisconsin-Madison 2002). Removal of the Elwha Dam was determined to have "major adverse short-term impacts on salmon" because sediment concentrations downstream of the dam would exceed lethal levels (National Park Service 1996).

- Release of toxic sediments. In an extreme and well-known case, removal of the deteriorating Fort Edward Dam on the Hudson River, New York in 1973 resulted in both biological and navigation impacts as several tons of PCB-contaminated sediments were released following removal (American Rivers et al. 1999).
- Release of nutrients. Gray and Ward (1982) found that the flushing of sediments from Guernsey Reservoir on the North Platte River caused a sixfold increase in downstream phosphorus concentrations, leading to the growth of large filamentous algal mats.
- Undesirable vegetation response. Though little empirical work has been done on vegetation and riparian response to removals, results may not be favorable and can include invasive exotics (Shafroth et al. 2002). Management is often necessary, though documented cases of unsuccessful active revegetation indicate that techniques need improvement (Drezner 2004).
- Physical instability and bank erosion (Figure 1.2). Depending on the volume and depth of stored sediments as well as project design choices, dam removals can lead to upstream channel incision, erosion and widening, with months and sometimes years of instability as the channel develops a new equilibrium. Transported sediments that deposit downstream can induce local and systemic instabilities if the depositional features create large islands or bars (Wohl and Cenderelli 2000).
- Risk of downstream ice damming. Dam removals in ice-prone rivers have been observed to cause increased risk

of ice jamming and damming (USACE 2001, White and Moore 2002). This should be compared to icing and ice transport dynamics of dam-affected downstream reaches. Incised and sometimes dewatered channels commonly observed below dams can produce large amounts of ice, which they are unable to route to floodplains during moderate flood events.

- Mobility of invasive organisms (e.g., sea lamprey). Dams serve as barriers to organism movement, which may prevent infestation from invasive species. Removal of the Marmot Dam in Oregon is being reconsidered, in part, because the dam not only restricts access by salmon to potential spawning sites, but also separates hatchery-reared and wild salmon, helping to maintain genetic integrity of the wild stocks (Oregon Department of Fish and Wildlife 2000).

Figure 1.2 Soudabscook River, ME channel instability after dam removal

Trade-off analyses must recognize that these costs and benefits are dynamic as knowledge grows and economies shift. Most of the potential ecosystem impacts of dam removal listed above can be eliminated or reduced when anticipated in effective project design and implementation using improved techniques. Few scientists and agencies recognized until recently that anadromous and adfluvial populations are critical sources of nutrients and trace minerals that support entire aquatic ecosystems. The former values of many dams have decreased, ceased, or been eclipsed by altered local and regional economies. In contrast, the role of dams in protecting infrastructure from flooding has increased with the widespread and often unmanaged development of floodplains.

Problem Analysis And Restoration Alternatives

Personnel addressing dam issues face an array of data demands that need clear and efficient articulation, development, consideration, and communication. For any given case, these can include:

Ecological

- Channel network analysis for habitat and life history needs in order to prioritize dam removal sequencing for species, community, and ecosystem restoration.
- Site, reach, and system effects of altered and restored sediment and hydrologic regimes.
- Threatened and endangered species, as well as other reference species or communities, e.g. recreationally or economically significant populations.
- Wetlands values with or without removal.
- Riparian characteristics and processes with and without removal.

Policy contexts

- Dam purpose(s) and condition
- Repair alternatives and costs

- Removal alternatives (full, partial, sequenced over time, and sequential grade control for headpond maintenance, organism passage, or channel stability) and associated costs
- In the event of continued operation or repair, dam maintenance needs
- Risk assessment of dam failure
- Dam mandates (e.g. licensing, fish passage) and liability
- Dam removal liability issues
- Dam and nearby property ownership and development
- Flood dynamics and mapping
- Flow rights and mandates
- Archaeological values
- Easements
- Public perceptions and stakeholder support for dams and dam removals
- Comprehensive cost-benefit analyses

Engineering and restoration implementation

- Sediment storage volumes, particle size analysis, contamination levels. Compare with upstream and downstream reaches
- Channel morphology downstream, through headpond, and upstream
- Icing and ice transport dynamics
- Hydrology and hydraulics
- Sediment transport capacity
- Bank stability
- Groundwater and well impact analyses
- Effects analyses on nearby infrastructure (water intakes, channel crossings, etc.)
- Restoration of upstream aggraded reaches and/or downstream incised reaches
- In the event of removal and the need for active sediment management, removal and disposal requirements (e.g., turbidity control, transport, disposal)

- Project sequencing and implementation
- Project effects monitoring (physical, hydrologic, biological, economic)
- In the event of dam repair or continued operations, determination of options for operational hydrology, fish passage, thermal regimes, riparian management, sediment amendments, or basin restoration for increased ecological integrity and productivity.

Regulatory Contexts And Agency Considerations

There are few policies to guide government agencies on dam removal considerations. Federal Energy Regulatory Commission (FERC), under pressure from Congress, developed a decommissioning policy in 1994, but it has been applied only once (on Edwards Dam in Maine). Doyle and Stanley (2003) suggested that dam removal is at a stage of "organizational learning" wherein agencies should maintain an adaptive position and act less as an advocate of favored solutions and more as a participant in deliberating problems and potential new solutions. The experience of Wisconsin as a center of dam removal activity offers valuable lessons (Born et al. 1998, Trout Unlimited and River Alliance of Wisconsin 2001).

The regulatory and management contexts of dams are onerous. The federal government is the largest single owner of dams at approximately 3 percent of inventoried structures; by contrast, 58 percent of ownership is private, according to the USACE National Inventory of Dams. Ownership of small dams changes hands and, over the years, dams are often abandoned to state dam safety agencies. Concerns regarding a legacy of potentially contaminated sediments associated with many small dams is a disincentive for many owners to become involved in decommissioning actions.

Federal agencies including the Corps of Engineers, FERC, U.S. Environmental Protection Agency (EPA), National Marine Fisheries Service (NMFS), USFWS, and the Natural Resources

Conservation Service (NRCS) all have an interest and role in dam removal actions, as do state and local governments and a wide variety of NGOs.

The diversity of interest, jurisdiction, and needs of these groups creates a multitude of opportunities for technical and funding collaboration.

The Corps is involved in dam removal issues through a variety of authorities, including Regulatory, small Continuing Authorities Projects, and Support for Others. The Corps is almost always involved in dam removals through its regulatory authorities, which address permitting under Section 404 of the Clean Water Act. The Corps' jurisdiction requires that a public interest review be carried out, as well as a determination of the effects of the dam removal on wetlands, fish and wildlife, water quality, water supply, energy conservation, navigation, economics, and historic, cultural, scenic, conservation, and recreational values. Environmental benefits and detriments and mitigation measures are also considered as part of the permit process.

Other regulations that must be addressed in a dam removal plan include the National Environmental Policy Act; the Fish and Wildlife Coordination Act; the Historical and Archeological Preservation Act; the National Historic Preservation Act; the Endangered Species Act; the Coastal Zone Management Act; the Marine Protection, Research and Sanctuaries Act of 1972 as amended; the Clean Water Act; the Archeological Resources Act; and the American Indian Religious Freedom Act.

Both the benefits and the impacts of dam removal actions transcend the boundaries of ownership, political jurisdiction and agency missions. Given the diverse purposes and objectives of the many entities involved in dam removal decisions, coupled with limited but emerging technical knowledge on the subject, it is not surprising that dam removal is a contentious subject guided primarily by opportunism rather than by scientific consensus or public policy.

Technical Guidance And Case Studies Analyses

Babbitt (2002) argued that, given the relative scientific ignorance of the consequences, current dam removal efforts are reminiscent of the dam-building era in the United States, and that we risk making decisions with similar costly long-term effects unless the ecological ramifications are better understood.

A literature of technical guidance for dam fate determination and removal is just emerging (ASCE 1997, USACE 2001, White and Moore 2002, this document). An early but excellent literature exists on social contexts and the process of dam removal (Aspen Institute 2002, American Rivers and Trout Unlimited 2006, Born et al. 1998, Heinz Center 2002). Useful listings and case history collections are also now available (American Rivers et al. 1999, Pohl 2001). The Water Resources Center Archives at University of California at Berkeley is creating a centralized clearinghouse for dam removal information and case studies.

The Engineer Research and Development Center (ERDC), U.S. Army Corps of Engineers, is currently planning and preparing a comprehensive list of publications that will assist practitioners in implementing dam removal projects. The ERDC is also initiating a three-year research effort to help overcome deficiencies in the state of knowledge regarding the removal of dams and the resultant ecosystem responses.

REFERENCES

American Institute of Biological Sciences. 2002. BioScience—A special section on dam removal and river restoration 52(8): 643-750.

American Rivers, Friends of the Earth, and Trout Unlimited. 1999. Dam removal success stories: Restoring rivers through selective removal of dams that don't make sense.

American Rivers. 2002. The ecology of dam removal—A summary of benefits and impacts. http://www.amrivers.org/damremovaltoolkit/eco logyofdamremoval.htm

American Rivers and Trout Unlimited. 2006. Exploring dam removal—A decision-making guide. http://www.amrivers.org/damremovaltoolkit/exp loringdamremoval. htm

American Society of Civil Engineers. 1997. Guidelines for retirement of dams and hydroelectric facilities. New York: ASCE.

American Society of Civil Engineers. 2002. Report card for America's infrastructure, 2001, sams. http://www.asce.org/reportcard.

Aspen Institute. 2002. Dam removal: A new option for a new century. Queenstown, MD: Aspen Institute. http://www.aspeninstitute.org/AspenInstitute/files/CCLIBRARYF ILES/FILENAME/0000000074/damremovaloption.pdf

Babbitt, B. 2002. What goes up, may come down. BioScience 52:656-658.

Bednarek, A. 2001. Undamming rivers: A review of the ecological impacts of dam removal. Environmental Management 27(6):803-814.

Born, S.M., K.D. Genskow, T.L. Filbert, N. Hernandez-Mora, M.L. Keefer, and K.A. White. 1998. Socioeconomic and institutional dimensions of dam removals: the Wisconsin experience. Environmental Management 22(3):359-370.

Collier, M., R.H. Webb, and J.C. Schmidt. 1996.

Dams and rivers—Primer on the downstream effects of dams. U.S. Geological Survey, Circular 1126.

Doyle, M.W., and E.H. Stanley. 2003. Toward policies and decision-making for dam removal. Environmental Management Vol. 31, No. 4, 453-465.

Drezner, T.D. 2004. Few native species colonize on mud flats ten years after dam removal (Wisconsin). Ecological Restoration 22(1):50-51.

Federal Emergency Management Agency. 2001. National Dam Safety Program. http://www.fema.gov/fima/damsafe/.

Gray, L.J., and J.V. Ward. 1982. Effects of sediment release from a reservoir on stream macroinvertebrates. Hydrobiologia 96:177-184.

Griffin, W. 2001. Frankfort, Winterport united over removal of dams. Bangor Daily News 10/29/01.

Heinz Center. 2002. Dam removal: Science and decision making. Washington, D.C.: H. John Heinz III Center for Science, Economics and the Environment.

Lary, S.J., and W.-D.N. Busch. 1997. American eel (Anguilla rostrata) in Lake Ontario and its tributaries: Distribution, abundance, essential habitat and restoration requirements. Amherst, NY: Lower Great lakes Fishery Resources Office, USFWS.

Magilligan, F.J., K.H. Nislow, and B.E. Graber, 2003. Scale-independent assessment of discharge reduction and riparian disconnectivity following flow regulation by dams. Geology 31(7):569-572.

National Parks Service. 1996. Elwha River Ecosystem Restoration Implementation Summary Draft EIS. Olympic National Park, Washington. http:/www.nps.gov/planning/olym/drftsum/elwha.htm

National Research Council (NRC). 1992. Restoration of aquatic ecosystems: Science, technology, and public policy. Washington, DC: National Academy Press.

Oregon Department of Fish and Wildlife (ODFW). 2000. http://www.dfw.state.or.us/public/newsarc/2000news/January/021000news.htm

Pohl, M. 2001. Constructing knowledge on American dam removals. In: The future of dams and their reservoirs. Denver, CO: U.S. Society on Dams.

River Alliance of Wisconsin/University of Wisconsin-Madison. 2002. Sediment Management Research Project, #FAF-0119. April, 2002.

Schuman, J.R. 1995. Environmental considerations for assessing dam removal alternatives for river restoration. Regulated Rivers 11:249-261.

Schuman, J.R. 2002. Environmental considerations for assessing decommissioning alternatives for large and small dams – sediment management. In Dams—Innovations for sustainable water resources. Denver, CO: U.S. Society on Dams, 485-496.

Shafroth, P.B., J.M Friedman, G.T. Auble, M.L. Scott, and J.H. Braatne. 2002. Potential responses of riparian vegetation to dam removal. BioScience 52(8): 703-712.

Trout Unlimited and River Alliance of Wisconsin. 2001. Restoring the flow: improving selective small dam removal understanding and practice in the Great Lake states. http://www.tu.org/pdf/newsstand/library/ restoring.pdf

U.S. Army Corps of Engineers (USACE). 2001. Considerations for dam removal in ice-affected rivers. Ice Engineering, #27. Cold Regions Research and Engineering Laboratory, U.S. Army Corps of Engineers.http://www.crrel.usace.army.mil/ierd/tectran/ 27In Design.pdf

White, K.D., and J.N. Moore. 2002. Impacts of dam removal on riverine ice regime. Journal of Cold Regions Engineering 16(1):3-16.

Wohl, E.E., and D.A. Cenderelli, 2000, Sediment deposition and transport patterns following a reservoir sediment release. Water Resources Research 36:319-333.

2

Funding Sources for Dam Removal[1]

Betsy Otto

Editor's Introduction: This paper is extracted from a guide book published by American Rivers explaining various sources of funds available for removal of dams. Many of these are Government organizations. These agencies contributed financially to dam removal because their objectives were served. We in India have similar government agencies. Our agencies do not appear to be aware that their objectives can be secured by prevention of dam building.

There is a need to be cautious, however. Most dams removed in the United States are out-of-use defunct dams. Funding was required for their removal often because the owners of these dams were not traceable and could not be held accountable for the environmental costs that the defunct dams

1. Adapted from Paying for Dam Removal, A Guide to Selected Funding Sources, American Rivers, 1025 Vermont Avenue NW, Suite 720, Washington, DC 20005.

were imposing on the society. The crucial learning for India is that the dams being built by us today may similarly become defunct in times to come and owners of these dams may not be around when these dams are ready for removal. There is a need to build-in costs of removal in all dam proposals. A National Fund for their removal must be established to which dams must contribute for their removal.

This paper is tells of the wide range of Government and non-Government agencies that are involved in dam removal. Our Ministry of Environment and Forests is parallel to the U.S. Environment Protection Agency. The EPA used its funds for Chesapeake Bay Program to pay for dam removal since that would allow fish migration. Our State Forest Departments have the mandate to preserve wildlife. This is parallel to the Pennsylvania Fish and Boat Commission which provided funds for dam removal.Our Pollution Control Boards are collecting penalties for water and air pollution. These monies should be used for the prevention of construction of new dams since the dams will add to pollution. The Connecticut Department of Environment Protection has done precisely this. State Governments in the U.S. have River Restoration Funds. This is parallel to our Ganga Action Plan. These funds must be used to remove barrages that cause river destruction. The National Fish and Wildlife Foundation of the U.S. has the mandate to develop fish resources. This agency is funding dam removal because it helps improve fish life. Such funding must be provided by our Ministry of Rural Development. Also, Ministry of Tourism may fund dam removal because free flow of river waters promotes tourism. The Ministry of Home should also fund such dam removal because it restores the right of pilgrims to bath and worship.

The average life expectancy of a dam is 50 years. Changes in land use and the structure of economy have reduced the need for certain dam functions. Our needs of hydropower may similarly change in future.

Introduction

Over the past 100 years, the United States led the world in dam building—blocking and harnessing rivers for a variety of purposes, including hydropower, irrigation, flood control, and water storage. The U.S. Army Corps of Engineers has catalogued approximately 75,000 dams greater than six feet along the waterways of the United States. Tens of thousands more small dams plug our rivers across the country.

While dams can benefit society, today science shows they also cause considerable harm to rivers. Dams change the chemical, physical, and biological processes of rivers all of which impact fish and wildlife. Dams block free-flowing river systems, hindering the flow of nutrients and sediments and impeding fish and wildlife migration. Dams also alter water temperatures and oxygen levels critical to good water quality and wildlife survival.

Many dams across the country have aged beyond their planned life expectancy, causing safety risks for communities downstream. The average life expectancy of a dam is 50 years, and a full one-quarter of all U.S. dams are now more than 50 years old. The American Society of Civil Engineers estimates that by the year 2020 that figure will reach 85 percent. In many cases, dam removal costs less than repairing an unsafe dam, especially where the benefits of the dam are marginal or non-existent.

Clearly dam removal is not appropriate for all—or even most—of the nation's dams. Many dams continue to serve public or private functions such as flood control, irrigation, and hydropower generation. This does not mean, however, that rivers should continue to be heavily impacted by these dams. Most dams across the country could be operated in a fashion that reduces their current negative impacts on the river. In hundreds of cases nationally, American Rivers and others are working to improve the operations of functional and economically viable hydropower dams through active participation in the federal licensing process. However, some dams cause such significant environmental damage that no amount of reoperation will alleviate the environmental harm. In many instances, dams no longer serve a purpose and may be abandoned. For these dams, where the environmental impacts of the dams outweigh the benefits, dam removal is often a reasonable and viable solution for restoring river functions.

Need for Dam Removal Funding

Removing dams for environmental benefits and to address unsafe and unwanted dams is still a relatively new phenomenon. Dam removals have been documented since the early 1900s—including a large number removed in just the last decade—and many more are undocumented. In part this reflects America's aging dam infrastructure; in part, it reflects significant changes in land uses and the structure of our economy, which has reduced our need for certain dam functions, as well as a growing concern about river ecology.

Many local communities, natural resource agencies, and environmental advocates want to remove selected dams that have outlived their purpose, are unsafe, or have costs that outweigh their benefits. The decision to remove a dam is often driven by safety concerns, but there may be compelling environmental and economic concerns as well. In many cases, dam removal saves significant taxpayer dollars compared to repair or environmental mitigation costs. On average, removal costs were only 37 percent of the estimated dam repair costs for 10 dams profiled in the report, *Dam Removal Success Stories: Restoring Rivers Through Selective Removal of Dams that Don't Make Sense.*

Finding funding for removal is a significant impediment to removing dams that don't make sense. There are almost no funding programs dedicated specifically to dam removal (Wisconsin is an exception). However, many federal, state and local government programs intended to improve water quality, protect or enhance wildlife habitat, restore natural resources or alleviate dam safety concerns can be used to finance dam removals. In addition, there are many sources of private funding, such as corporate environmental damage mitigation funds (these funds may be government-administered) that can be used to remove dams. For example, dams in Maine, Wisconsin, Pennsylvania, and other states have been successfully removed using creative approaches that combine multiple types of public and private financing. The share of various sources of funding in dam removals in the United States is shown in Fig 2.1. It is seen that State Governments and Corporate sectors have the largest contributions.

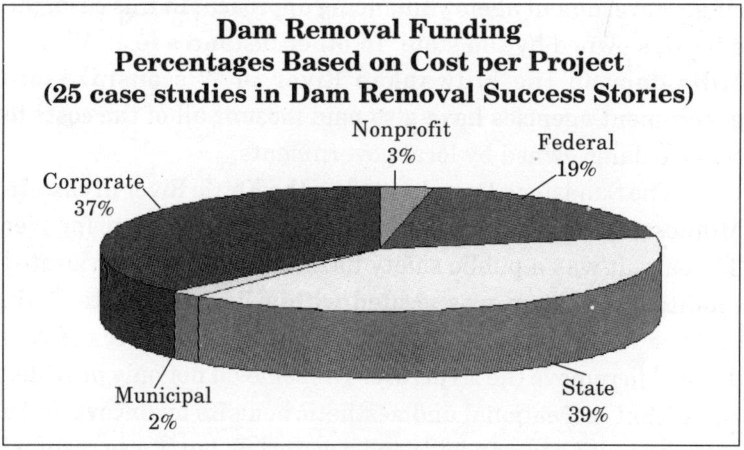

Fig. 2.1

The information provided in this report is cause for both concern and optimism. The lack of dedicated funds for dam removal foretells an increasing problem as dams across the country age and the need for investment in repair and removal becomes more critical. It also exposes the potential for a significant lost opportunity. As we better understand the negative impacts that dams have on rivers, fish and wildlife, and water quality, removal of dams that don't make sense can be a simple, cost-effective way to alleviate many of the problems associated with dams. It would be very unfortunate and short-sighted to miss these restoration opportunities simply because of the lack of funds for dam removal.

Examples of Funded Projects

Three examples drawn from the Dam Removal Success report illustrate some typical dam removal financing approaches that bear out these figures and trends.

Single Funding Source: Sandstone Dam, Kettle River, Minnesota

The Sandstone dam removal offers a typical example of a single government agency financing approach. In this case, the dam was owned by the state. In other instances (e.g., Woolen Mills dam on the Milwaukee River in Wisconsin) state government agencies have also paid most or all of the costs to remove dams owned by local governments.

The Sandstone Dam, located on the Kettle River in eastern Minnesota, was an inactive hydropower dam. Obsolete for over 30 years, it was a public safety hazard due to its deteriorated condition. The dam was located within Banning State Park, and the Minnesota Department of Natural Resources (DNR) decided to remove the structure. The removal not only provided significant recreational and aesthetic benefits by uncovering a stretch of whitewater rapids and a waterfall, but it also restored fish habitat for numerous species, including the rare lake sturgeon. Minnesota Power and Light gifted the structure and 200 acres of surrounding land to the DNR when the cost of producing power became uneconomical in the 1960s. When the dam was removed in 1995, the cost of refurbishing the dam for hydropower was estimated at over $1 million compared to an estimated removal cost of $300,000. The actual cost of removal was only $208,000. Funding for the project, as well as engineering support, were accomplished with the agency staff and budget of the Dam Safety Program at the Minnesota DNR, Division of

Waters. A private company was employed to conduct the actual demolition of the 20-foot tall and 150-foot wide structure. Due to limited funding, little stream restoration was done in conjunction with the project other than some initial bank stabilization.

Cooperating Agencies: Seven dams on the Conestoga River, Pennsylvania

Removal of the seven dams from the Conestoga River and its tributaries in southeastern Pennsylvania provides a good example of a federal and state government agency cooperating to share costs and achieve complementary objectives—improvement of an migratory fishery and enhancement of river health. All were obsolete run-of-the-river dams that were originally built to power mills or provide navigation canals with water. The dams on the Conestoga, a large tributary of the Susquehanna River, blocked migratory fish, including American shad, from reaching their historic spawning grounds.

The Pennsylvania Fish and Boat Commission managed the removal of the dams, which ranged from $1,500 to $110,000 in cost, and from three to 13 feet in height and 10 to 300 feet in length. Half of the $218,500 removal costs were funded through the U.S. EPA's Chesapeake Bay Program for migratory fish passage. This program requires a 50 percent match from a non-federal source, which was chiefly provided by the Pennsylvania Fish and Boat Commission. Other local government agencies and non-governmental groups provided in-kind services to assist with the removals and contribute to the 50 percent cost-share.

Multiple Sources: Seven dams on the Naugatuck River, Connecticut

The dam removals on the Naugatuck River provide an excellent example of bringing together multiple sources of funding to make a project possible.

The Connecticut Department of Environmental Protection (DEP) has been working to restore the degraded Naugatuck River under a far-reaching program that includes full removal of four dams—the Anaconda and Freight St. Dams in Waterbury and the Plats Mill and Union City Dams in Naugatuck—as well as construction of fish and/or boat passage at three others, and major upgrades to six wastewater treatment plants. DEP has worked in partnership with local communities, Trout Unlimited, and other private partners to accomplish the river-wide restoration and to arrange an estimated $8 million in total funding to complete the work on all seven dams.

An elaborate funding package was put together to make the project work, including the following components:

- To pay for the overall dam removal planning and design, Connecticut DEP used approximately $300,000 in supplemental environmental penalties—in this case, payments in lieu of environmental enforcement penalties from Clean Water Act and State Clean Water Act violations.
- The City of Waterbury accessed approximately $300,000 in Clean Water Act funds for upgrading its wastewater treatment plants. As mitigation for water quality violations during construction to expand its facility, the City eventually provided $1 million for the removal of Platts Mill and Freight St. Dams as well as fish passage improvements to other dams.
- Connecticut DEP removed Union City Dam for $250,000 using state River Restoration Fund monies, which are financed through state bonds.
- An additional $50,000 for the Union City Dam came from the Iroquois Pipeline Fund (on a 5:1 matching basis) through the National Fish and Wildlife Foundation, a quasi-federal government funding source for resource conservation and restoration. (An additional $50,000 in Iroquois Pipeline monies will go toward the fish and boat bypass planned for the Tingue Dam on the Naugatuck.)

- Finally, Anaconda Dam was removed by the city of Waterbury with $100,000 of its own funds, although the city is suing the dam owner to recoup these costs.

PART-II

EXAMPLES OF DAM REMOVALS IN UNITED STATES

3

Economic Analysis of Removing Four Dams on the Lower Snake River[1]

Taxpayers for Common Sense and others

Editor's Introduction: This paper is extracted from a detailed Cost-Benefit Analysis of four lower Snake River dams in Washington state. Our practice is to make a single Cost-Benefit Analysis (CBA) of the proposed dam. This paper shows this is not correct. The correct method is to make two analyses—with and without dam.

The Columbia and Snake River salmon plan proposed to raise populations of salmon fish. The paper points out that the costs of doing this while maintaining the dams were underestimated. Similarly, the costs of prevention of fish

1. Adapted from "Revenue Stream: An Economic Analysis of the Costs and Benefits of Removing the Four Dams on the Lower Snake River."

migration due to construction of dams were underestimated. This criticism would be applicable to, for example, to the prevention of migration of Mahseer fish in Uttarakhand. Removal of dams will improve water quality of the river and save funds required for recovery plans. Similarly, not making dams will save funds required by Ganga Action Plan.

Dams trap huge amounts of sediments in the reservoirs. This leads to raising the reservoir water level by a few inches every year. This sediment build-up has elevated the water level behind a levee in Lewiston, Idaho. As of 2002, the 5 feet of levee protection had been reduced to only 2 feet. High river flows would put Lewiston at risk of flooding. This cost is not internalized in our EIAS.

Run-of-river dams do not generate electricity when power is most in demand. Our ROR dams similarly provide most power during monsoon season when the demand is less. Unfortunately, this low value of power is ignored by CEA while the high value during peak hours is recognized. Thus we adopt lopsided valuation of power.

Removal of the lower Snake River dams and replacing their power with clean energy would not impede economic growth, and could create as many as 15,000 long term jobs. The idea that dams are required for 'economic growth' and will provide jobs to people has to be re-examined in the light of this finding. Removing the four lower Snake River dams will draw hikers, birders, boaters, campers, and other visitors from across the nation and around the world. Interest in the restored river will generate diverse recreation-based economic activity.

Revenue Stream demonstrates that removing the four lower Snake River dams in Washington state will return significant economic dividends to the Northwest and the nation, creating a "revenue stream" of both cost savings and economic benefits from new and restored industries. Removing these four dams — four out of more than 220 major dams in the Columbia

River Basin — will likely save American taxpayers and Northwest electricity ratepayers a minimum of about $12 million to $2 billion over 10 years, and $2 billion to $5 billion over 20 years, and restore Snake River salmon in the process. Beyond the billions saved, removing the four lower Snake River dams will ultimately produce as much as a five-fold increase — between almost $4 billion to $7 billion over 10 years — in new annual revenue generated from tourism, recovered fish runs and outdoor recreation.

Between 1961 and 1975, the U.S. Army Corps of Engineers constructed four federal dams on the lower Snake River. The dams — Lower Granite, Little Goose, Lower Monumental, and Ice Harbor — were built primarily to facilitate barge travel 140 miles upriver to what became the inland port of Lewiston, Idaho.

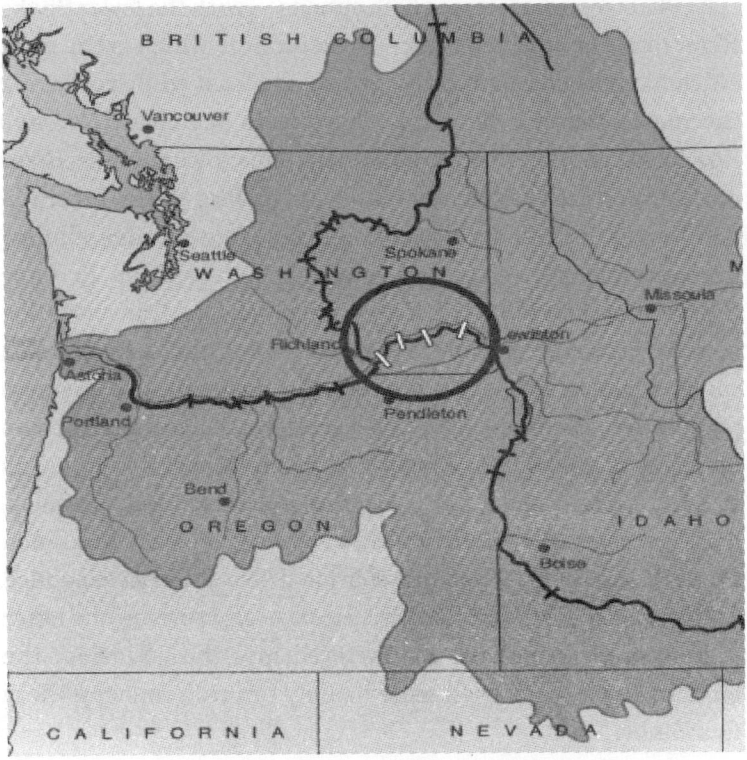

They also generate a small percentage of the Northwest's electricity. While these four dams provide some benefits, they cost taxpayers and Northwest electricity ratepayers hundreds of millions of dollars each year and are jeopardizing the survival of wild Columbia and Snake River salmon and steelhead. Federal agencies have identified the federal dams as the major cause of decline in these salmon populations and have discussed whether it is necessary to remove the four lower Snake River dams in order to meet federal laws and treaties.

This report answers the following questions:

- What are the costs of restoring salmon with and without the dams?
- What are the economic benefits with and without the dams?

This report examines both options — with the lower Snake River dams or without them — to determine which is the most efficient choice economically, costing the least while returning the most economic benefits. The results are clear: *Revenue Stream* concludes that retaining the four lower Snake River dams would cost taxpayers billions of dollars in the future in addition to the billions already wasted during decades of failed attempts to recover Columbia and Snake River salmon. *Revenue Stream* also shows that the energy, irrigation, and transportation benefits currently provided by the four lower Snake River dams can be replaced efficiently and cost-effectively. From a business perspective, the financial costs of keeping and maintaining these dams far outweigh their benefits. It will be cheaper for taxpayers to remove these dams than to continue their annual subsidies, and dam removal is a cost-effective way to restore wild salmon to the Snake River. When the additional benefits from expanded commercial and recreational fishing, boating, tourism, and other recreational industries are factored into the equation, the balance sheet leans even more heavily toward removing these four dams.

Comparing Federal Expenses and Savings With and Without the Four Lower Snake River Dams

The costs to taxpayers and Northwest ratepayers were estimated for two scenarios:

1. **The Costs of Keeping the Four Lower Snake River Dams.**

This scenario examines the costs of attempting to recover Snake River salmon with the lower Snake River dams in place, plus other expenditures that will be required to keep the dams operational and to raise local levees to keep Lewiston, Idaho from flooding. Many of these expenses are currently being paid annually; others — including additional salmon recovery costs, sediment control, and dam repairs — are necessary future costs in addition to existing spending. The total costs of keeping the dams in place range from $7.8 billion to $9.1 billion over a 10-year period and from $15.7 billion to $18.2 billion over a 20-year period.

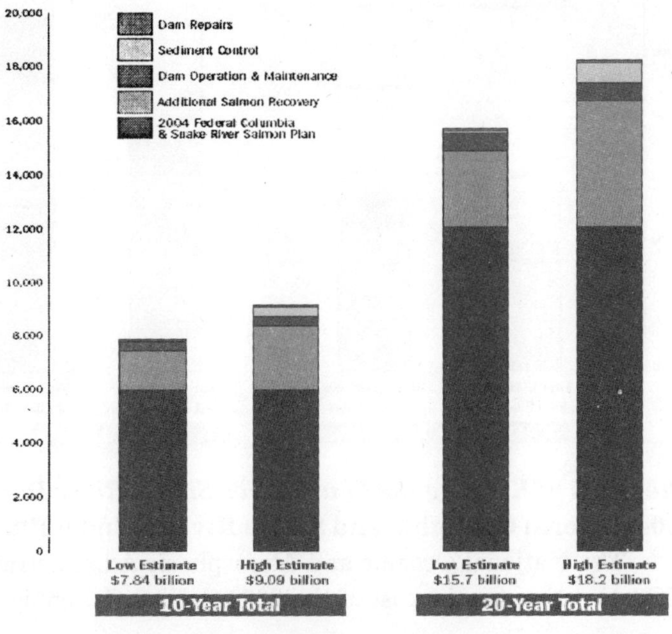

2. The Costs of Removing the Four Lower Snake River Dams.

This scenario examines the costs of physically removing the four lower Snake River dams, costs associated with replacing the benefits that the dams currently provide, and costs required for salmon restoration still needed after dam removal. The total costs of removing the dams and replacing the benefits that the dams currently provide range from $6.2 billion to $9.1 billion over a 10-year period and from $11.1 billion to $16.6 billion over a 20-year period.

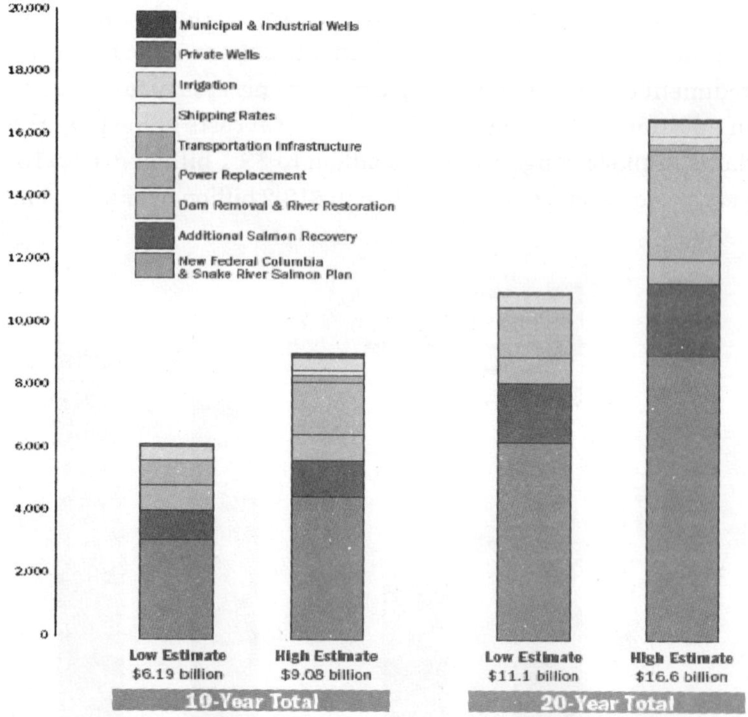

The Costs of Keeping the Four Lower Snake River Dams

2004 Federal Columbia and Snake River Salmon Plan

The National Oceanic and Atmospheric Administration (NOAA) estimates the cost of the 2004 federal Columbia and

Snake River salmon plan to be approximately $6 billion over its 10-year life, or about $600 million per year. Despite its hefty price tag, the 2004 plan allows salmon to decline below today's already depressed numbers. In May 2005, a federal court ruled this plan — also known as a "biological opinion" or "BiOp" — illegal. As a result, this cost likely represents a conservative estimate of the cost of a legally valid plan that leaves the four lower Snake River dams in place, assuming such a plan could be developed.

Additional Salmon Recovery Costs with Snake River Dams

Federal law requires the federal government not only to prevent extinction as described in the previous section, but also to take steps to recover salmon and steelhead to a point where they no longer need federal protections. In an attempt to meet its salmon recovery obligations, the federal government has undertaken a "recovery planning" process that, when complete, will outline recovery criteria and actions over and above those required by the $6 billion federal salmon plan. The Yakama Indian Nation recently estimated that these additional recovery costs could total between $1.42 billion and $2.34 billion over 10 years, or about $142 million to $234 million per year.

Dam Operation and Maintenance

Continuing to maintain and operate the four lower Snake River dams requires significant annual federal expenditures. According to the Army Corps, taxpayers and Northwest electric ratepayers pay about $33.7 million annually to operate and maintain the dams. This includes payments to keep the dams' navigation locks and power generation facilities in working order. These costs will continue throughout the lives of the dams, and include no salmon-related costs.

Sediment Control with Snake River Dams

Lower Granite Dam, the farthest upstream of the four lower Snake River dams, traps roughly 3.2 million cubic yards of sediment annually, which raises its reservoir water level by approximately 3 inches per year. Since the dam was completed in 1975, this sediment build-up has elevated the water level behind a levee in Lewiston, Idaho. As of 2002, the designated 5 feet of levee protection had been reduced to only 2 feet.

The more sediment that collects behind Lower Granite Dam, the more danger this situation poses to the community of Lewiston. The Army Corps now estimates that significant rainfall during spring runoff or other periods of high river flows would put Lewiston at risk of flooding. Simply put, Lower Granite Dam increases Lewiston's flood risk.

Major Dam Repairs

The turbines in all four lower Snake River dams were put into operation between 1961 and 1979, and have approximate lifespans of 25-50 years. Therefore, keeping them operational will require significant and immediate work, such as repairing and upgrading the turbines and turbine blades and rewinding the generators. The Army Corps estimates that over 40 years (10 years per dam, one dam at a time) this work will cost roughly $225 million, or $5.6 million per year.

The Costs of Removing the Four Lower Snake River Dams

New Federal Columbia and Snake River Salmon Plan

With dam removal, federal taxpayers and Northwest ratepayers would benefit from substantial savings in salmon recovery costs. According to a study compiled for the Northwest Power and Conservation Council, removing the lower Snake River dams would reduce the cost of Columbia and Snake River Basin salmon restoration by 35 to 55 percent — requiring just $3.1 billion to $4.5 billion over 10 years. Over 20 years, the costs would likely double.

Additional Salmon Recovery Costs

Removing the four lower Snake River dams would decrease Columbia River Basin-wide implementation costs for federally required recovery plans by about 45 percent, to between $95.6 million and $116 million per year. Such reductions are possible because most tributary habitat in the Snake River Basin is in good condition, and the benefits from removing the lower Snake River dams would reduce the need for expensive tributary habitat improvements in this part of the Columbia River Basin. Remaining funds could be directed to other areas of the basin, primarily Columbia River tributaries outside the Snake River Basin, in order to maximize the positive benefit of habitat restoration investments.

Dam Removal and River Restoration

The Army Corps estimates the one-time cost of removing the earthen portions of the four lower Snake River dams at approximately $790.5 million. The Army Corps analyzed removing only the earthen portion of each dam.

In addition to the physical removal of each earthen portion, the Army Corps' cost estimate includes: channelization to control the undammed river's course as reservoir levels are lowered; revegetation of the newly exposed land; and protecting highway and railroad bridges that cross the reservoirs as well as the adjoining roads and railroad tracks, among other actions.

Replacing the Benefits of the Four Lower Snake River Dams

Power Replacement

Combined, the four lower Snake River dams generate about 1,200 average megawatts (aMW) annually, representing less than 5 percent of the entire Pacific Northwest's average energy consumption. However, because these are "run-of-the-river" dams with almost no ability to store water in order to generate electricity when power is most in demand, the power they generate is of low value to the Northwest. In the winter and late summer, when energy demand is higher and surplus power most valuable, river flows are at their lowest and the combined output of the four dams meets only about 2 percent of Northwest energy demand.

The Northwest's highly variable precipitation pattern limits the dependability of these dams. Bonneville Power Administration (BPA), which markets the hydropower from these dams, puts their annual total firm power — the amount that can be counted on in a drought — at only 790 aMW. In December, for example, when the region needs it most, the four dams together produce only 423 aMW of firm power. Consequently, the dams are relatively unreliable for serving electricity loads. That unreliability means that the region must keep expensive, gas-fired combustion turbines on hand— but underutilized — for times when power from the dams is unavailable. This drives up the costs of these four dams.

In 2002, the RAND Corporation studied the economic impact of removing the lower Snake River dams and replacing their power with clean energy (energy efficiency and wind power). The results showed that doing so would not impede economic growth, and could create as many as 15,000 long term jobs.

A NW Energy Coalition analysis estimates that replacing the power from the dams with a mix of energy efficiency and wind power would increase electrical utility costs $79 million to $170 million per year for the next 20 years. Actual increases in monthly energy bills would be modest —about 65 cents per month for residential customers if spread among the millions of Pacific Northwest ratepayers, or about $2 per month if paid only by BPA's firm customers.

Transportation Infrastructure Investments: Converting from Barge to Rail

Prior to the completion of the lower Snake River dams in 1975, grain and other products in the region were transported to market chiefly by rail and truck. Today, a significant portion of these products move via barge from Lewiston, Idaho, or grain-loading facilities elsewhere on the lower Snake River. Recent studies have found that the 140-mile navigation channel created by the lower Snake River dams could be affordably and effectively replaced by upgrading the Northwest's railroad lines. Upgrading railroads in southeastern Washington and Idaho to accommodate most of the grain currently moving down the lower Snake River (some would still be barged from Columbia River ports near Pasco, Washington) would cost between $17.7 million and $230.6 million. The large range in this estimate reflects varying assumptions about the ability of existing rail-served grain elevators to handle higher volumes of grain after the lower Snake dams are removed.

Increased Shipping Rates

Burlington Northern Santa Fe Railway predicts that rail shipping rates will be "fully competitive" with existing barge rates after the lower Snake River dams are removed. Other observers predict a modest increase in shipping rates. A Northwest transportation consulting firm estimates that annual shipping costs for all goods and commodities currently shipped down the lower Snake River will increase $7.1 million to $13.1 million per year after lower Snake River dam removal. This translates into an increase of about 3 to 7 cents per bushel of grain, which comprises more than 85 percent of all Snake River barge shipments.

Irrigation Investments

Ice Harbor Dam is the only one of the four lower Snake River dams that provides irrigation for farms. According to the Army Corps, a one-time investment of about $421 million would allow the 13 affected farms, irrigating a total of 37,000 acres, to continue irrigating with Snake River water after the lower Snake dams are removed.

The Army Corps' estimate is based on consolidating all Ice Harbor irrigation-pumping stations into one primary station and distribution system. The $421 million includes the one-time cost of constructing the pumping station and distribution system as well as its annual operation and management costs for the following 10 years. To date, no independent study has been performed to determine if more cost-effective irrigation modification options exist.

Private Well Modifications

Removing the four lower Snake River dams would lower the water table near the river and could require modifying privately owned wells located within one mile of the lower Snake River. The modifications, which would involve deepening 71 wells and improving water-pumping systems, would require a one-time expenditure estimated at $65.5 million.

Municipal and Industrial Water Use Modifications

Removing the lower Snake River dams would require modifications to some municipal and industrial pumps. The Army Corps estimates the one-time cost of these modifications, which would be necessary at only one of the four lower Snake River reservoirs, at $13.3 million to $63.9 million. Well over 90 percent of these costs (or all but $700,000) would go toward modifying the water-cooling system at a Potlatch Corporation pulp mill in Lewiston, Idaho, to ensure its compliance with the Clean Water Act. Depending on the cooling system built at the mill, the estimated cost would be either $12.5 million or $63.2 million. The remaining municipal and industrial pump changes would go toward two Public Utility District pumps used by two local golf courses and a concrete aggregate washing company.

Summary: Comparing the Scenarios

Taxpayers and Northwest electricity ratepayers could save almost $5 billion, and restore Columbia and Snake River salmon in the process, by removing the four lower Snake River dams and completing the other tasks necessary for overall recovery. This includes the cost of replacing the current benefits of those dams; it does not include the costs or benefits associated with job gains and losses.

The status quo has already cost taxpayers more than $7 billion, and has produced few tangible results. In summary, taxpayers could save approximately $2 billion to $5 billion over a 20-year period.

Regional Economic Benefits of Lower Snake River Dam Removal

Lower Snake River dam removal will bring considerable savings to American taxpayers and Northwest ratepayers. It will also provide significant local and regional economic benefits. Removing these four dams will create a "revenue stream" of expanded sport and commercial salmon industries, restore 140 miles of river for recreation activities, and unlock the major earnings potential of the Snake River Basin.

Non-fishing Recreation

Removing the four lower Snake River dams will draw hikers, birders, boaters, campers, and other visitors from across the nation and around the world. Interest in the restored river will generate diverse recreation-based economic activity, especially in inland Northwest communities from Lewiston, Idaho, to Pasco, Washington, and many spots in between. An economic study commissioned by the Army Corps concluded that removing the lower Snake River dams would generate between $223 million and $360 million per year in non-fishing recreation. That is nearly ten times the $36.6 million per year in economic activity now generated by reservoir-based recreation on the lower Snake River.

Fishing: Recreational and Commercial

Removal of the four lower Snake River dams and subsequent recovery of Snake River salmon and steelhead populations will increase fishing opportunities throughout the Columbia and Snake River Basin and in five states: California, Oregon, Washington, Idaho, and Alaska. Recreational and commercial salmon fishing opportunities in the Columbia and Snake rivers and Pacific Ocean currently are limited in order to protect imperiled populations, such as those returning to the Snake River. The restoration of wild Snake River salmon and steelhead will open new fishing opportunities not only for currently protected species, but also for healthy salmon populations. However, given available information, accurately predicting the economic value of fishing-related benefits is a difficult task. A more comprehensive and consistent economic analysis is necessary to confidently assess dam removal's full economic benefit for recreational and commercial fisheries. The following are estimates based on the best available information to identify the economic activity from expanded commercial and recreational fishing after removing the four lower Snake River dams.

Sportfishing

Recreational fishing for salmon and steelhead in the Columbia and Snake River Basin currently supports a vital economy in the Northwest. Recreational fishing involves spending on boats and related fishing gear (rods, reels, tackle, etc.), as well as travel expenses (food, gas, lodging, etc.) in and around river and coastal communities. This economic output ripples through local communities, helping to foster jobs, economic stability, and growth.

A 2005 economic study finds that fully restored salmon and steelhead fishing in Idaho alone could provide economic activity in excess of $556 million annually statewide. Much of that activity ($300 million) would be realized in small rural communities. This total is an increase of almost 300 percent over the economic activity Idaho experienced from salmon fishing in 2001 and steelhead fishing in 1992-93.

Since it does not account for Oregon and Washington economic activity, the $556 million per year estimate for Idaho can be viewed as a conservative estimate of sportfishing economic activity throughout the Northwest after dam removal. That still represents $470 million more each year than the Northwest states generate with the four lower Snake River dams in place.

Commercial Fishing

Available data indicate that Snake River salmon restoration will bring considerable economic benefits to commercial fishing communities. A recent study by an economic advisory board to the Northwest Power and Conservation Council finds that with the four lower Snake River dams in place, the average value of commercial fishing for Columbia and Snake River salmon over the past 30 years has been $55.7 million per year. The study includes no estimate of the additional value of restored Columbia and Snake salmon and steelhead runs.

A report by the Institute for Fisheries Resources estimates that recovery of all Columbia River Basin stocks to historic levels could generate $637 million per year for the commercial fishing industry over and above current levels. According to commercial fishing industry experts, at least one-fifth of the increased economic activity — that is, $127.4 million per year — would come from restoring Snake River stocks alone.

Conclusion

Federal laws and treaties require the protection and restoration of wild Columbia and Snake River salmon. Two options — retaining four dams on the lower Snake River in Washington state and removing those same dams — are now under discussion, each with its own costs and benefits. A side-by-side comparison of the options shows that removing the four lower Snake River dams is less costly for taxpayers, and generates greater economic benefits. Restoring wild salmon with the lower Snake River dams in place would cost taxpayers $7.8 billion to $9.1 billion over 10 years, and generate non-fishing recreation and sport and commercial fishing benefits of $1.8

billion to $6.5 billion over the same time period. Conversely, restoring wild salmon by removing these four dams should be the cheaper of the two options for taxpayers, costing between $6.2 billion and $9.08 billion over 10 years. That is $12 million to $1.7 billion less than the costs of keeping the dams in place. Additionally, removing the dams results in economic activity from non-fishing recreation and sport and commercial fishing equaling as much as almost five times today's benefits (with the dams in place), ranging from $9.1 billion to $10.5 billion over 10 years.

Removing the dams saves both taxpayer and Northwest electricity ratepayer dollars and at the same time provides significant new revenue to the Northwest. Keeping the dams in place is clearly the more expensive option and brings the fewest benefits while falling far short of guaranteeing restoration of Columbia and Snake River Basin salmon. The expenses associated with retaining the four lower Snake River dams could accrue without actually meeting the requirements of federal law. Removing the dams is less expensive and is more certain to meet federal law. *Revenue Stream* also identifies areas in which additional independent economic assessments of the true costs and benefits of dam removal are needed. In particular, economic analyses of the potential benefits of a restored fishery in the Columbia River Basin are sorely needed. *Revenue Stream* shows the need for Congress to demand and ensure the completion of these studies.

Without independent analyses, it is likely that the nation will continue to waste precious resources – both economic and natural. *Revenue Stream* demonstrates that removing the four lower Snake River dams is a serious salmon-recovery option that would both save money and provide economic benefits. Consequently, national and Northwest leaders should closely examine this option rather than continuing to spend taxpayer dollars on less-beneficial measures. Taxpayers and Northwest electricity ratepayers deserve this thorough examination.

Total Federal Savings Over Ten Years

Costs	*Low Case*	*High Case*
With Four Lower Snake River Dams	$7,835,867,382	$9,093,026,064
Without Four Lower Snake River Dams	$6,185,846,545	$9,081,145,776
Federal Savings Without Dams	$1,650,020,837	$11,880,288

Total Regional Economic Benefits Over Ten Years, Without Lower Snake River Dams

Benefits	*Low Case*	*High Case*
Non-fishing Recreation	$2,232,908,459	$3,600,231,750
Sportfishing	$5,585,215,606	$5,585,215,606
Commercial Fishing	$1,273,885,350	$1,273,885,350
10-Year Total Benefits	$9,092,009,415	$10,459,332,706

4

Non-use Values Lead to the Removal of Elwha Dam[1]

Battelle

Editor's Introduction: This study was undertaken to assess the costs and benefits of retaining or removing the Glines and Elhwa dams on Elhwa River. The first EIS was completed in 1995. It determined that removing both the Glines and Elhwa dams would be better than installation of fish ladders.

The impact of sediment flows during removal was examined. One alternative was to dredge and transport the sediment through pipelines to the sea. This indicates the severity of the problem of dislodging the accumulated sediments behind dams. These are never considered in our Environment Impact Assessments.

1. Adapted from "Economic Support for the Elwha River Watershed: Final Economic Characterization Report with Monitoring Recommendations" prepared for The Coastal Services Center, National Oceanic and Atmospheric Administration, May 21, 2007.

The 1995 report was not acted upon till 2002. In this period many changes took place. These included the closing down of one of the industrial water customers, the listing of two additional fish species as threatened, growth in population, and changes in the feasibility of recycling concrete from the Glines Dam. The import is that change in economic circumstances necessitates a new EIA. In India, however, a EIA is never revised. It is assumed to be valid in perpetuity.

The supplemental EIS suggested, among others, to create a water treatment facility to remove sediment during the sediment erosion, raise the Federal levee and complete other flood protection measures, create an accessible hiking trail for observation of the dam removal and site, and purchase space to offset the impact of the loss of the reservoirs for the trumpeter swans. This last observation is important. It requires the project to buy land to mitigate adverse environmental impacts. This is almost never considered in our EIAs. Our EIAs state that land will not be acquired even for making Green Belt around the reservoir rim. The Green Belt is often left to be made voluntarily by the landowners on the reservoir rim.

The benefits from removal included the positive impacts on the fisheries, the positive impact on shellfish, the positive impacts on recreation and tourism, the cost savings associated with reduced need to carry out maintenance at Ediz Hook, and the non-use or existence values associated with dam removal and river restoration (the value people place on knowing a specific natural setting is in good condition and available to future generations even though they do not intend to experience it first hand).

The benefits to the Elwha Klallam Tribe were assessed at two levels. One, those related to fishing. Two, those relating to cultural heritage. The report says: the benefits from fishing do not capture the enormous benefits to the Tribe, who place tremendous cultural and spiritual importance on the river, its fish, and the historic sites inundated by the dams' reservoirs.

This observation is relevant for us. All our rivers are worshipped as deities. This worshipful attitude will certainly be hit by diverting the river into tunnels and canals. This enormous loss is almost never examined in our EIAs. This is specially relevant to the dams being built on the Ganga, which has been declared to be our National River.

Overall, the findings were that benefits significantly outweighed the costs for removal of both dams. The result was largely driven on the benefits side by the inclusion of the non-use values. Hindus from all parts of the country similarly have a huge non-use value for the Ganga which is not internalized in cost-benefit analysis of dams being built on this holy river.

Introduction

In preparation of the planned 2009 removal of two dams on the Elwha River located on the Olympic Peninsula in Northwest Washington, the National Oceanic and Atmospheric Administration (NOAA) Coastal Services Center (the Center) is working with the University of Idaho and the Lower Elwha Klallam Tribe to develop baseline data and a characterization of the lower Elwha River in order to support the long-term management of the resource by the Tribe and other stakeholders. As part of this effort, NOAA has contracted Battelle, through Photo Science, Inc., to carry out the economic research described in this report.

The purpose of the Battelle work is the characterization of the economic impacts associated with Elwha River dam removal and subsequent river restoration.

There are two hydroelectric dams on the Elwha River: the 108-foot Elwha Dam, built 4.9 miles from the mouth of the river in 1911, and the 210-foot Glines Dam, built 13.4 miles from the river's mouth in 1925. The reservoir created by the Elwha Dam is known as Lake Alwell, and the reservoir created by the Glines Dam is Lake Mills. The dams restrict the indigenous anadromous fish populations to the lower 4.9 miles of the river, though historically they had access to the full 45 miles of the Elwha River and 100 miles of tributary streams now restricted by the dams. This is to the obvious detriment of the fish, but also to the detriment of the complex ecosystem the fish previous helped support, and the American Indian population that relied on the fish populations economically and spiritually. Under the Treaty of Point No Point signed in 1855, the Elwha Klallam ceded their native title to the land in exchange for the promise by the Federal Government of the United States of permanent rights to fishing, hunting, and gathering in their traditional areas. The construction of the Elwha Dam, just 56 years later, was a private effort on privately owned land but decimated the Klallam's ability to fish the river. The dam also failed during construction, sending forth flooding that damaged Indian homesteads.

In 1992, Congress passed the Elwha River Ecosystem and Fisheries Restoration Act (Public Law 102-495). This initiated the river restoration project and all of the study and planning documented in a series of environmental impact statements (EIS) prepared by the NPS in coordination with the U.S. Fish and Wildlife Service, Bureau of Reclamation, Bureau of Indian Affairs, U.S. Army Corps of Engineers, and the Lower Elwha Klallam Tribe.

The first EIS was completed in 1995 and evaluated the different alternatives for restoring the fisheries and ecosystem as mandated by Congress. The alternatives included installing fish passage measures on the dams, removing only the Elwha Dam and installing passage measures on the Glines Dam, removing only the Glines Dam and installing passage measures on the Elwha Dam, and the preferred alternative, which was removal of both dams. These were each evaluated for impact relative to the "no action" case under which both dams remained without intervention to improve fish passage. The EIS identifies the relevant fish species, their estimated recovery rates for each alternative, as well as the impacts on other aquatic and non-aquatic wildlife and vegetation. These impacts are considered not only in the context of the improved habitat that the majority of species are expected to experience after restoration, but also with respect to expected negative impacts associated with the construction activities—release of sediment and the loss of the habitat afforded by the reservoirs. Additionally, the EIS addresses impacts to safety, flooding, cultural resources, land use and recreation, and water quality. Water quality is a highly significant issue for the project because of the pent-up sediment retained behind the dams. The 1995 EIS deferred evaluation of alternatives for dealing with sediment to an implementation EIS to be completed after one of the overarching alternatives for restoration was chosen. The alternative chosen was the removal of both dams.

The implementation EIS addressed alternatives for the implementation of the removal of both dams. Some of the sediment removal and other alternatives were ruled out in initial review and described, but not extensively analyzed in the EIS. One alternative that was analyzed in the EIS but not chosen was to use a dredge and slurry technique, together with a pipeline to transport sediments to the Strait of San Juan De Fuca. This was expected to reduce the impact of sediment on water quality during the dam removal process. The preferred alternative, which was chosen, was to allow the river to naturally erode the sediments, with the rate of erosion controlled by the rate of the dam removal. The implementation EIS also covers the mitigation efforts that will be undertaken to maintain flood risks at current levels, preserve water quality and availability for a variety of water users, and mitigate the impacts of rising ground water on septic systems.

A notice was issued in 2002 of the intention to prepare a supplemental EIS because the continued study of the restoration project mitigation alternatives, along with local project-related changes, had led to the conclusion that different alternatives would be preferable to those described in the 1996 EIS. The changes since 1996 include the closing down of one of the industrial water customers, the listing of two additional fish species as threatened, growth in population, and changes in the feasibility of recycling concrete from the Glines Dam. The supplemental EIS was completed in 2005. These changes led to the decision to create a water treatment facility to remove sediment during the sediment erosion and a new permanent treatment facility, pursue surface water rather than an infiltration gallery, protect the Dry Creek Water Association's well field from flooding, connect the Elwha Heights Water Association to the Dry Creek Water Association, relocate the Tribal fish hatchery with the assurance that it has an adequate water supply, continue operation and assure water supply for the State fish rearing channel, raise the Federal levee and complete other flood protection measures, create a waste water collection and treatment system for the reservation, create an accessible hiking trail for observation of the dam removal and site, and purchase space to offset the impact of the loss of the reservoirs for the trumpeter swans. The 2005 supplemental EIS also includes changes to the vegetation plan.

During the decision-making process while reviewing the alternatives for the first EIS for the Elwha River restoration, the Elwha Human Effects Team (the Team) carried out an economic analysis of the anticipated effects of the project, titled *Economic Analysis Final Technical Report* [1995]. The Elwha Human Effects Team was made up of analysts representing the Lower Elwha Klallam Tribe, NPS, and the U.S. Bureau of Reclamation. The analysis was carried out for a 100-year period, beginning with the anticipated initiation of dam removal and projected into the future. This time frame was the maximum duration of a study allowable by the Water Resource Council's *Principles and Guidelines*, an important source of guidance on economic analysis of water related projects.

The Team used a benefit cost analysis (BCA) framework for carrying out the analysis. In this type of economic analysis, the positive features of each alternative (the benefits) are carefully tabulated in conjunction with each alternative's negative features (the costs). The sources of benefit the Team identified included the positive impacts on the fisheries, the positive impact on shellfish, the positive impacts on recreation and tourism, the cost savings associated with reduced need to carry out maintenance at Ediz Hook, and the non-use or existence values associated with dam removal and river restoration (the value people place on knowing a specific natural setting is in good condition and available to future generations even though they do not intend to experience it first hand).

The methodology the Team used for calculating the benefits associated with fishery recovery was based on biological projections of the recovery of fish stocks, the schedule for allowed harvests, and the distribution of the harvests among sports fishermen, commercial Tribal fisheries (under treaty), and non-tribal commercial fisheries. The increase in value added that the catches represent to the commercial industries associated with fishing (fishers, processors, retail) all were calculated using assumptions for market harvest prices and price mark-ups at each stage in the production process. The value for sports fishermen was calculated based on anticipated participation and sports fisherman expenditures per day drawn from a survey.

Non-fishing recreation and tourism benefits were based on a survey that estimated the potential increase in visitation after the restoration is complete, as well as data that described the duration of trips, the type of lodging preferred (camping versus hotel), and typical spending per day, all assessed in the context of existing lodging capacity.

Ediz Hook, which provides import protection to the harbor at Port Angeles, is expected to benefit from the removal of the dams because sediment will be allowed to flow freely down the river rather than being impounded by dams. This would reduce the current requirement to replenish Ediz Hook artificially. The Team used the actual costs of the existing replenishment program to value the reduction in replenishment costs.

Non-use values were assessed by the Team based on a survey carried out to estimate how highly residents of Clallam County, the rest of Washington State, and the rest of the United States value the resource affected. This survey was carried out using a technique called contingent valuation (CV). This was the only non-market source of benefit considered in the economic analysis. When tabulated, it produced significantly larger values than any of the other sources of benefit.

The Team also assessed the magnitude of the economic impact associated with the direct project spending that would take place within Clallam County. This is not included in the BCA, but represented an important distributional feature of the project. Another important distributional feature of the project is how it will benefit the regional Native American population beyond the Tribal Treaty commercial fishery income, in particular the Lower Elwha Klallam Tribe. The benefits to the Tribe associated with the fisheries are relatively easy to measure, but do not capture the enormity of the benefit to the Tribe, who place tremendous cultural and spiritual importance on the river, its fish, and the historic sites inundated by the dams' reservoirs. The Team described the adverse economic conditions of the Tribe relative to the general Clallam County population and Washington State population, characterized the importance of the river and fish to the Tribe, and detailed the history of the Tribe's interaction with the U.S. government and American settlers including the building of the dams.

The costs identified in the Team's analysis include the direct construction costs, the loss of electrical power generation, and costs of flood mitigation (which were included within the construction costs). The cost of mitigation efforts in the implementation and supplemental EISs for local municipal and industrial water, septic systems, and flooding, were not known to the Team. Like the original flood mitigation costs, these additional project mitigation cots would be considered direct costs of the dam removal and river restoration project categorically considered with the construction costs. The Team relied on estimates of these costs for other project documentation that were generally engineering oriented.

Overall, the findings were that benefits significantly outweighed the costs for removal of both dams over the range of discount rates used to estimate net present values. This was not the case for any of the other alternatives. The result was largely driven on the benefits side by the inclusion of the non-use values.

5

Removal of Edwards Dam[1]

Michael Paulson

Editor's Introduction: This news report was published immediately after the Edwards dam in Maine was removed. The dam was required by the Federal Energy Regulation Commission, which is equivalent of our Central Electricity Authority, to install fish ladders to enable salmon and other fish to migrate upstream to their spawning areas. The owner of Edwards Dam, facing a $9 million price tag to build a fish ladder, agreed to allow the dam to be removed. And the owners of other upstream dams, as well as Bath Iron Works, agreed to pay the $7 million cost of removal as mitigation for the harm their own projects do to fish.

This narrative indicates that economics of a dam is deeply connected with its environmental impacts. In India, however,

---

1. Adapted from "One dam down; others in line," Seattle Post-Intelligencer, Seattle, Friday, July 2, 1999.

the CEA approves hydropower projects without examining the economic costs of mitigating environmental impacts. The CEA only provides that the promoters will obtain approval from the Ministry of Environment and Forests. It is implicitly assumed that the environmental mitigation measures will not impact the economic viability of the project.

This narrative of the Edwards dam disproves this assumption. It shows that environmental mitigation measures can decisively impact the economic viability of a hydropower project. One reason this happens in India is that adverse environmental impacts are grossly understated leading to nominal costs of mitigation.

This narrative points to the need of a unified authority for economic and environmental clearances to hydropower projects in the country. The present model of two separate, independent and disjoint approvals from CEA and MOEF is fundamentally flawed because it fails to appreciate the interrelation of environmental mitigation and economic viability.

This report also tells of the live debate between hydropower companies and environmentalists. It indicates that the ability to decommission dams rests with the vociferousness that environmentalists display in opposing the dams. It is not that the FERC has become anti-dam. Truly, FERC is pushed to make stipulations that lead to removal of dams. Indian environmentalists need to adopt a similar strategy. They can tip the scales against making of dams by documenting the adverse impacts of dams and seeking their true mitigation.

Yesterday, on an overcast summer morning in the lush green Kennebec Valley, a lone backhoe scooped about 30 buckets of earth from a temporary gravel wall that had been built to patch a 60-foot gap engineers had cut in Edwards Dam.

Suddenly, accompanied by the pealing of church bells and cheers of hundreds of spectators, the caged Kennebec River turned wild.

Carrying more than a century of sediment and seven generations of memories, an estimated 18,000 cubic feet of water per second roared through the breach and, for the first time since 1837, flowed freely toward Merrymeeting Bay and the ocean 43 miles away. Soon, nine species of fish, including the endangered shortnose sturgeon, will once again be free to populate the river.

Fig. 5.1. Workers watch as water flows through during breaching of the 162-year-old Edwards Dam on the Kennebec River in Maine. AP photo

Environmentalists and anglers plan to use their success here to renew their push to remove more of the nation's estimated 75,000 dams. And at the top of their wish list are eight in Washington state.

"This is the beginning of something that is going to affect this entire nation," said Interior Secretary Bruce Babbitt, who is pictured on his official biography holding a sledgehammer.

He referred to the removal of the Edwards Dam as "a statement about our capacity to honor and respect God's creation, the sacramental commons, and to live not just in the past, but in a visionary and a different future in a way of balance and harmony with creation."

Babbitt's enthusiasm for dam removal horrifies defenders of some of the Northwest dams, which kill salmon but do generate hydropower and support irrigation, navigation and recreation.

"This (dam removal) is a further indication that Secretary Babbitt has this in mind and takes pleasure out of doing these things," said Rep. George Nethercutt, R-Wash., a fervent opponent of removing any dams from the Snake River in Washington state. "It's a perverse sort of pleasure."

The breaching of 24-foot-high, 917-foot-long Edwards Dam marks the first time a federal agency has ordered a dam removed after deciding that the cost to migrating fish denied access to upstream spawning grounds was greater than the benefit from the hydropower.

Similar decisions could be reached in Washington state, whose growth is partly attributable to the economic benefits of hydropower. At the same time, salmon, the very symbol of the region, are struggling to survive because dams block some of their migratory routes.

The most controversial proposal in the state is a push by environmentalists, driven by a desire to restore endangered salmon stocks, to remove four 100-foot-high federal dams on the lower Snake River. Ice Harbor, Lower Monumental, Little Goose and Lower Granite dams are valuable for irrigation and hydropower but harmful to fish.

"Late this year or early next year, we will come out with a recommendation either to breach or not to breach, or some third way," said Terry Garcia, the assistant secretary of commerce responsible for the National Marine Fisheries Service. "We're going to start engaging the region in a wide-ranging debate, not just on the science, but on all of the components to a decision that would affect those dams and that river . . . and hopefully reach a consensus."

But consensus is not in sight. Pitted against the environmentalists are powerful economic interests opposed to dam removal. Elected officials from the state have lined up to denounce the concept.

"This debate is irrelevant, when we ought to be saying many species have been listed as endangered, and what do we need to do to get them unlisted," said Sen. Patty Murray, D-Wash. "We're spending a lot of time debating whether the Snake dams should come down, and not enough time putting our energy into other ways to save salmon."

A more likely candidate for removal in the short term is the Elwha Dam on the Olympic Peninsula. Congress has voted to remove the dam, and is considering how to pay for it.

But there are also several other, lesser-known instances in which a dam could be removed much sooner. The Simpson Timber Co. in Shelton hopes next summer to remove an unused 14-foot dam from Goldsborough Creek, a south Puget Sound tributary, while PacifiCorps, a Portland-based utility, is negotiating with environmentalists over how to pay for the likely removal of Condit Dam on the White Salmon River in Southwest Washington.

Some environmentalists are pushing for the removal of the Enloe Dam on the Similkameen River in Okanogan County, while others think Cushman Dam, owned by Tacoma Public Utilities, should go.

People have been building dams for thousands of years — the first known dam was built on the Nile by the Egyptians in 2800 BC. Early American colonists used dams to power grist mills and saw mills, building their first dam in South Berwick, Maine, in 1634.

But removing them has become a trend. An estimated 121 dams have been removed in the United States in the past 40 years, but the breaching of the Edwards Dam "really marks a turning point in how we view rivers across the country," said Margaret Bowman, senior director of dam programs at the environmental group American Rivers.

"In the past, dam removal was considered a very radical and not very feasible way of restoring a river, but the Edwards demonstrates that this can be a practical and cost-effective way of repairing damage."

The future of dams has become an international issue. The World Commission on Dams is studying a variety of projects, including Washington state's Grand Coulee Dam, in an effort to assess the impacts of dams.

The hydropower industry is bracing for the debate. Industry officials are urging policy-makers and consumers to consider the benefits of hydropower.

"Over the next 10 or so years, the lion's share of hydro dams are coming up for relicensing, so this is really a pivotal time for hydropower," said Richard McMahon, an official of Edison Electric Institute, an industry trade association. "A lot of these dams are extremely productive resources — emissions free, renewable — and we'd like people to see the bigger picture."

In Maine, the picture was considerably less cloudy than in the Northwest.

By the time the dam removal movement was in full swing, the Edwards was generating just a tenth of a percent of Maine's electrical power, and the mills for which the dam was built had long since closed.

In 1997, the Federal Energy Regulatory Commission denied a new license to Edwards, citing environmental costs. Fish once so plentiful that a laborer's contract stipulated he would not be served salmon more than once daily disappeared from above the dam and were struggling to reproduce below it.

The owner of Edwards Dam, facing a $9 million price tag to build a fish ladder, agreed to allow the dam to be removed. And the owners of other upstream dams, as well as Bath Iron Works, agreed to pay the $7 million cost of removal as mitigation for the harm their own projects do to fish.

Yesterday, the transformation of the river was dramatic. Above the dam, what had been a giant muddy reservoir populated by worms and freshwater fish became a rushing river that dropped 10 feet and soon is expected to attract the salmon and sturgeon that populate the lower river, as well as predatory birds such as osprey and eagles. Below the dam, the blue-gray water turned brown as the sediment and debris began to flow — a temporary disturbance from which the river is expected to recover.

Federal authorities predict that removing the dam will add 1,100 shortnosed sturgeon to the river and double the number of Atlantic sturgeon to 10,000 over the next three to five years, as well as benefiting other migratory fish, including the American shad, Atlantic salmon, striped bass, blueback herring and alewives.

The cost of the dam to fish has been seen around Memorial Day each year, when hundreds of thousands of fish gather below the dam, flopping around after failing to traverse the dam to reach their upstream spawning habitat. Next year, they are expected to swim as far in as Waterville, 17 miles north, until they hit the next dam.

6

Removal of Embrey Dam[1]

Hal Wiggins

Editor's Introduction: This tale of removal of the Embrey Dam on the Rappahannock River in Virginia was written by an officer of US Army Corps of Engineers. The lessons for India are as follows.

Change in economic circumstances led to the dam becoming economically unviable. Construction of large nuclear plants led to low price of electricity and rendered small hydropower plants as the Embrey Dam unviable. The hydropower plants being built by us may get similarly outdated with cheap solar power on the horizon and mega thermal power plants in the making.

Alternative non-intrusive ways of withdrawing river water for drinking purposes can be made. The Embrey Dam was providing drinking water to the city of Fredericksburg through

1. Adapted from *A Tale of Two Dams: From Salem Church Dam to the Embrey Dam* by Hal Wiggins, USACE, Black Cat Press, 2006.

a canal. It blocked the flow of Rappahannock River wholly and diverted water into a canal. This function was now done by the Motts Run pumping facility (details at next chapter). This facility pumps out water from the river during high flows without obstructing the main flow. Dams like Bhakra, Sardar Sarovar and Tehri block the flow of the river like was being done by Embrey Dam. We can replacc these dams with a facility that withdraws and stores water during high flows without blocking main flow of the rivers.

Dams develop cracks in course of time. They are not 'permanent' structures. The Embrey Dam was constructed in 1910. It developed cracks and stared leaking and became unsafe in 1990s. Our economic assessments of dams do not include this cost.

Fish passage was an important aspect of dam removal. The State Government commissioned a study on providing fish passage. This report recommended removal of the dam as the best option. The same logic applies to worship of the rivers in India. Removal of dams will provide benefits to pilgrims. It is remarkable that we in India do not place value on the right of pilgrims to worship the rivers, which is protected by the Constitution, while our fellow human beings in United States place value on the right of fish to migrate.

Efforts of individual legislators bear fruit. Removal of the Embrey Dam was triggered by State Senator Edd Houck who secured a resolution from the State Senate requesting the Virginia Department of Game and Inland Fisheries to conduct a study on providing fish passage at Embery Dam. Again, through the efforts of Senator John Warner of Virginia, congressional authorization was written to provide the federal government with 100 percent of the funding to remove the Embrey Dam and crib dam. Our Members of Parliament can similarly work for removal of specific dams.

Removal of sediment from behind the dam was a major issue. The silt, gravel, clay and water deposited behind the dam was propelled through a 12-inch- diameter flexible pipeline to a 13-acre dredge containment area. We will face similar costs when we have to decommission our dams. These costs must be built into the economic assessments.

Religious sanction for letting the rivers flow. During the dam removal, Senator John Warner quoted a passage from Ezekiel in the Bible "There will be a very great multitude of fish, because there waters go there; for they will be healed and everything will live wherever the river goes." Our scriptures have much more to say of the spiritual qualities of river waters. We need to listen to these.

Amid great applause and ceremony in Fredericksburg, Virginia on February 23, 2004 the Embery Dam was rocked by two explosion that opened the river. A wave of water was seen charging down the river carrying debris and a flock of hungry sea gulls hovering above the swirling waters. The breaching of this 94-years old dam was a great event that allowed the Rappahannock River to run free for the first time since 1854.

I became intimately aware of the Embrey Dam and fledgling movement to remove the dam to support the cause of migratory fish. One of my first memories of this advocacy was when I was introduced to Larry Gross, a local river peddler and enthusiast. As we shook hands, Larry accused me of being one of those "dam builders" with the Corps of Engineers. I immediately felt that there was some history with the locals and my employer, the U.S. Army Corps of Engineers. Indeed, there were echoes of the past still lingering with the Salem Church Dam, a juggernaut civil-works project planned thirty-five years ago that would have dammed up the up the Rappahannock River along a twenty-mile river segment. This dam didn't happen. I didn't know it at the time, but I and many others would soon be involved in a landmark event in the planning for the removal of the Embrey Dam. Public values in regards to the environment would dramatically change over a period of 30 years. The era of great government funded civil works projects such as dam building was over. A new era was dawning where the environmental concerns of the public championed the cause of free-flowing rivers. I felt that this story needed to be told.

Figure 6.1: The Embrey Dam in operation

The Embery Dam was constructed from 1909 to 1910 by the Fredericksburg water power company. The dam stood 22 feet high and 1,070 feet long including abutments on each side. In conjunction with the 1854 wooden crib dam, it was the only mainstream impediment on the Rappahannock River to anadromous or spawning fish. These two dams blocked an estimated 700-1000 mainstream and tributary miles of historic spawning habitat. Constructed of reinforced concrete, Embery Dam was an early example of an Ambursen dam, a slab built into the Embery Dam on the Southern abutment. The ladder was quiet steep and was never know to move one fish above the dam without human invention. The construction of the dam also included the construction of the new hydroelectric plant, and represented an important period in the evolution of hydroelectricity in the Fredericksburg region. The dam and the production of the energy allowed the mills and other industries to be built away from waterways where they were in danger of flooding. This had a positive impact on the region's development.

In 1994, an engineering analysis of the dam paid for by the City noted excessive cracking in the dam's concrete, with deep pockets and trenches, plus significant leakage. The report stated that the continual flow of water through these fissures would accelerate the deterioration of the concrete and eventually breach the dam. The dam was henceforth seen as maintenance and safety liability by the City. In 1987, a tragic event occurred when a teenager fell off the dam and was killed. As a result of this incident and others, the City of Fredericksburg began in 1995 to investigate various alternatives to address the safety and environmental issues connected with the dam in 1995. The City commissioned a study by Whitman, Requardt and Associates to determine the relative costs associated with three choices: 1. Partial breach of the dam, 2. Fish ladder for migratory fish and 3. Making the Embrey dam a permanent gravity mass dam. This last alternative was actually considered as an option by the City. However, due to public advocacy, the City began at that time to seriously consider a full or partial breach of the dam.

There were several obstacles as well as sound arguments against the removal of the Embrey Dam. The citizen effort, sponsored primarily by FOR (Friends of Rappahannock), the City of Fredericksburg and other local and state efforts such as the Virginia Department of Game and Inland Fisheries, garnered support to remove the Embrey Dam. This idea initially met with opposition since the City's water supply was associated with the Embrey Dam, as the dam provided a backwater pool for an intake on the river. In addition to this, a breach in the Embery Dam would result in the loss of flowing water into the Rappahannock Canal. Since no river water would flow into the Rappahannock Canal once the dam was gone, the Crops offered to pump water into the canal from its lower end at the tidal section and to build a $500,000 air diffusion system to keep storm water that runs into the canal oxygenated and healthy.

One by one, these obstacles had to be removed in order to develop a plan for the removal of the dam. On May 17, 1996, Senator John Warner met with officials on Trench Hill in Fredericksburg to broker an agreement to move ahead with federal involvement with the Embrey Dam project. It was decided that a federal/state sharing of the costs for the removal of the dam would be pursued. In 1997, the City of Fredericksburg began making plans to abandon its water supply intake at the Embrey Dam.

A new regional Motts Run Raw Water Treatment Plant was built in 1998 that pulls water out of the Rappahannock approximately 5.6 miles upstream of Fredericksburg. This major governmental action removed a major obstacle to the removal of Embrey Dam since the public water supply intake at the dam was abandoned. Now no one could say that the dam helped provide water for the City of Fredericksburg.

It was realized that the breaching of the dam would result in a sudden and uncontrolled release of sediment into the Rappahannock River. The concerns over the amount and quality of the sediment behind the dam would need comprehensive study. Over the century and a half, hundreds of thousands of cubic yards of sediment had accumulated behind the Embery Dam. What was in the sediment? How would the sediment migrate downriver and disrupt aquatic life and navigation? A comprehensive study was to be conducted to sample and test the sediment behind the dam and to identity an appropriate dredge containment facility to receive the dredge material prior to breaching the dam. The City of Fredericksburg was no longer opposed to the removal of the Embrey Dam, however, a condition of the removal of the dam would be to provide and maintain water in the City's Rappahannock Canal. The Embrey Dam itself and the crib dam behind it were also considered to be historic. The City of Fredericksburg eventually has to purchase land from the Silver Companies, a local developer in town in order to provide the U.S. Crops of Engineers with a dredge containment facility for the sediment behind Embrey Dam.

The Commonwealth of Virginia also had a strong interest in the removal of the Embrey Dam. Primarily through the efforts of State Senator Edd Houck, a Virginia State Senate Joint Resolution No- 296 was passed which requested the Virginia Department of Game and Inland Fisheries to conduct a study on providing fish passage at Embery Dam. On December 9, 1997 the Embery Dam Fish Passage Study was released by Timmons, Inc. It outlined alternatives and issues to be evaluated in providing fish passage at Embery Dam. The findings of the Timmons report stated that " the recommended option for providing fish passage at Embrey dam is to remove all or part of the dam." Timmons estimated that there was over 3000,000 cubic yards of sediment behind the dam that may have to be removed. The cost calculated for the complete removal of the dam was $ 7,450,000.00. It was also determined in the Timmons report that filling the Embrey Dam with concrete and making it a permanent gravity mass dam with fish passage would cost $10,208,750,00. Due to the costs involved, and the complexity of environmental and cultural issues, the City agreed to partner with the federal and state governments to bring down the dam.

The U.S. Army Corps of Engineers was tasked by Congress in 1999 to evaluate the removal of the Embrey Dam in Cooperation with the City of Fredericksburg, Stafford County and the Commonwealth of Virginia (VDGID).

The compelling rationale behind the removal of the Embrey Dam was fish passage. Tidal rivers in Virginia and throughout the Chesapeake Bay have historically supported large populations of migratory fish which travel up rivers in the spring to spawn. The anadromous fish species of interest that are known to spawn in the Rappahannock River such as

American shad, alewife, blueback herring, hickory shad and striped bass traditionally support major commercial and recreational fisheries. These fish species also provide an important ecological function as a food source linking the marine, tidal estuarine and nontidal food chains. Since the 1970s, these species have experienced severe declines in abundance. Since 1991 in the Rappahannock River, shad stocks were at a very low abundance and declining; alewife were at a low level of abundance but stable; blueback herring were moderately abundant and stable; and hickory shad existed as a remnant population. According to the U.S. Fish and Wildlife Service, the reasons attributed to this decline include over Fishing and loss of spawning and nursery habitats due to pollution and impediments such as dams.

In 2001, the U.S. Congress appropriated funds for the Preconstruction Engineering and Design for the removal of the Embrey Dam. The Norfolk District, Army Corps of Engineers thus formed a Project Delivery Team to develop a comprehensive plan for the demolition and removal of the Embrey Dam from the Rappahannock River. The comprehensive analysis involved feasibility studies, engineering design plans for the dredge containment area, wetland delineations, environmental studies, cultural resources studies and mitigation plans. These studies needed to be done to address the short-term and long-term impacts of the dam removal. In 2002 and 2003 a large crater-shaped dredge containment facility had to be constructed on the south side of the Rappahannock River. In August 2003, work began as over 250,000 cubic yards of dredge material was hydraulically dredged from the Rappahannock River behind the Embrey Dam. The silt, gravel, clay and water sucked from being Embrey Dam would flow through a 12-inch- diameter flexible pipeline to a 400-horsepower booster pump on shore that propels the mixture uphill to the 13-acre dredge containment area. The whole line would run more than 3,000 feet. Due to concerns about additional sediment remaining in the river upstream of

Embrey Dam after this initial dredging, an additional 21,000 cubic yards of sediment was removed in early 2005.

The preparation for the explosive breach in the historic Embrey Dam involved six hundred pounds of explosive set by the Army and Air Force Reserves. These explosive were placed in a 130-foot section near the middle of the dam. The Corps worked with the U.S. Army Dive Company and the U.S. Air Force 555th RED HOESE Reserve Squadron on the demolition under the Department of Defense's Innovative Readiness Training Program to carry out the mission.

The big event had finally arrived. Speeches were given by dignitaries of local, state and federal government. In the words of Tayloe Murphy, the secretary of Natural Resources for the Commonwealth of Virginia, "A new day is beginning for this river. Anglers will flock here to fish for shad, herring, and striped bass. Watermen will work their nets downstream. Canoeists and kayakers will ply these waters from the mountains to the Bay." Senator John Warner quoted a passage from Ezekiel in the Bible "There will be a very great multitude of fish, because there waters go there; for they will be healed and everything will live wherever the river goes." Warner took a ceremonial sledgehammer carved with shad running along its handle and the words "SHAD LIBERATOR" and began the countdown to demolition. Local folk singer and guitar maker Bob Grumman sang his tune he composed in 1992, titled "Rappahannock Running Free." The song mentions Embrey Dam and as Grumman sang to the top of his lungs, "I say it's time we blow it up, there's no need for this dam," the crowed went crazy with applause and cheers.

It took two blasts to open up the Embrey Dam. The first blast opened up only about 10 percent of the dam. This first blast rocked the area and provided a show that brought a wall of water down the river up the bank full mark along the shore.

There were set-backs during the work in the river to remove the dam. In October 2003, an 8,000 pound dredge barge

Figure 6.2: Rappahannock River flowing free near the Embrey dam site.

broke loose from a cable during a high water event and washed the barge over the Embrey Dam. During January 2005, the causeway and most of the large metal pipes washed downriver as the river crested at twelve feet in Fredericksburg, carrying them downstream and scattering them along both sides of the river. Most of these pipes were quickly recovered from the freezing water by local paddlers in the area. Finally, on March 15, 2005, the last remnants of the Embrey Dam, crib dam causeway and all construction debris were removed from the Rappahannock River. A group of five recreational paddlers were seen that day, floating by and cheering enthusiastically.

7

Denial for Wilkes-Barre Inflatable Structure[1]

U.S. Army Corps of Engineers

Editor's Introduction: This paper is extracted from the order of U.S. Army denying the permission to build an inflatable dam on the Susquehanna River for flood control and enhancement of recreational facilities. The order determined that the economic value from the proposed dam is less than the economic value that can be created from riverfront development. This is the true meaning of 'alternatives' that are required to be considered as a part of Environment Impact Assessments under the EPA in India. The USACE held that the objective of enhancing recreational facilities can be accomplished through a number of non-structural alternatives which do not obstruct flow of the river. It took upon itself to examine whether the

1. Adapted from Department of the Army Environment Assessment and Statement of Finding, Document No CENAB-OP-RPA (Wilkes-Barre Inflatable Structure) 2003-01240-13.

stated objective necessarily required making a dam. In India the regulatory agencies like Central Electricity Authority should likewise make an assessment whether hydropower is the only alternative to production of power.

It is customary in EIAs done in India to list endangered species of flora and fauna. If none are present then it is assumed that a project will not have any negative impact on environment. A similar situation prevailed in the Wilkes-Barre proposal. No federally listed or proposed threatened or endangered species were known to occur within the project area. But the U.S. Fish and Wildlife Service did not stop here. It pointed out many other impacts: such as obstruction to fish migration, deterioration of water quality, alteration of riparian wetlands and shallow water habitats and sediment deposition and erosion. Our State Forest and Wildlife Departments should likewise go beyond endangered species and look at the impact on wildlife more holistically.

This document also brings out a lively debate between the Applicant City and US Fish and Wildlife Service (USFWS). The USACE sits in judgment over the submissions made by USFWS and the replies given by the Applicant City. We in India have no such procedure. The Applicant Company itself provides the EIA and MOEF determines whether the proposed mitigation measures are adequate. The proceeding is between two parties with no window for one who wishes to oppose the project. The approvals given by the Expert Appraisal Committee of MOEF to River Valley Projects are non-speaking. There is no requirement for them to reply point-wise to the objections raised even in public hearings. There is a need for us to modify our procedures and to make it obligatory for the sanctioning authority to (1) hear the opposing parties; and (2) to give point-wise replies to the objections raised.

The applicant is proposing to discharge dredged and fill material into the Susquehanna River, and adjacent wetlands, for the construction of an inflatable structure that will span the

Susquehanna River. The structure will be located on the river bottom from the south bank of the Susquehanna River. The applicant is proposing to inflate the structure from approximately Memorial Day through Labor Day every year to create a seasonal recreational impoundment that will extend approximately 4.5 miles upstream from the structure.

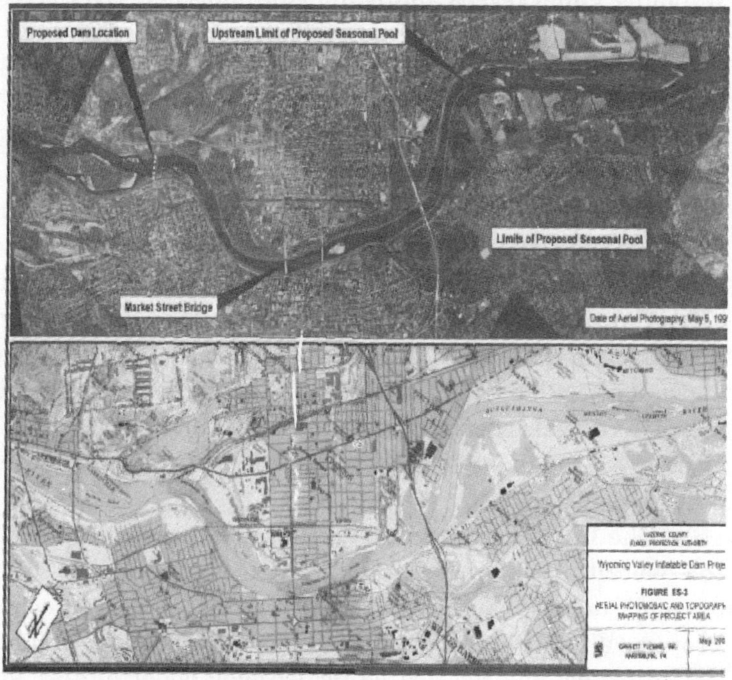

Fig. 7.1. Project Location and Vicinity Map

A Denil fish passage facility is proposed along the north bank of the river as well as an eel ladder. The applicant also proposed to use an "operational flexibility" design to mitigate water quality concerns. Operational flexibility is the term used to describe potential adjustments to the inflatable bladders that can raise or lower the height of the structure. These adjustment could be used to lower the structure during high flow events or to mitigate water quality concerns associated with Combined

Sewer Overflows (CSOs) and to allow for a flow-through system to ensure water flow quantity and velocity are maintained downstream.

The basic project purpose for the Wilkes-Barre Inflatable Structure (WBIS) is to "mitigate for impacts to river access and loss of recreational opportunity caused by the raising of the Wyoming Valley levees." Mitigation for impacts to river access and loss of recreational activity can be accomplished through a number of structural or non-structural alternatives which do not require siting in special aquatic sites. For example, such alternatives or combination of alternatives could include portals for access between downtown Wilkes-Barre and the riverfront, fishing piers with public access, annual sponsored recreational fishing/canoeing/Kayaking tournaments, walkways, interpretive trails, and other strategies that would enhance recreational opportunities in the Wilkes-Barre area. Therefore, I have determined that this basic project purpose is not water dependent, and in accordance with the Guidelines, practicable alternatives which do not involve discharges into special aquatic sites are presumed to exist unless clearly demonstrated otherwise.

The Corps has determined that the three objectives of the overall project purpose: mitigation for intangible socio-economic impacts caused by the Wyoming Valley Levee Raising Project (WVLRP), economic development for the Wyoming Valley region, and river based recreational opportunities, are integrated and necessary to meet the applicant's purpose and need and to allow for proper consideration and evaluation of alternatives. The applicant's needs have been considered in the context of the desired geographic area of the development and the type of project being proposed. The geographic scope of the alternatives considered are limited to locations that would mitigate for adverse impacts to the Wyoming Valley at Wilkes-Barre and include alternatives that are structural and non-structural to meet the three objectives of the overall project purpose statement.

For purposes of a permit application the Corps advised the applicant that their analysis of practicable alternatives must include non-impoundment alternatives. In addition, on May 22, 2007, the Corps assisted the applicant in developing the six alternatives that are evaluated in this document. These six alternatives comprise the appropriate range of practicable alternatives considered within the context of the 404(b)(1) alternatives analysis and NEPA review. The result of this analysis will determine whether or not the applicant's preferred alternative is the light of the basic and overall purpose, in accordance will the Section 404(b)(1) Guidelines.

In addition to the no action alternative, two non-structural alternatives and three structural alternatives were evaluated in the context of the 404(b)(1) analysis as well as the public interest review. The following alternatives were carried through for assessment in this document:

(1) No Action Alternative.
(2) Riverfront Development Plan.
(3a) Inflatable Structure with Denil Fish Way.
(3b) Inflatable Structure with Vertical slot fish way.
(3c) Inflatable Structure with operational flexibility.
(4) Inflatable Structure with Denil fish way plus Riverfront Development Plan.
(5) Inflatable Structure with Denil Fish way plus Riverfront Development Plan plus West Bank Riverfront Development Plan (applicant's preferred alternative).
(6) Riverfront Development Plant plus West Bank Riverfront Development.

Potential aquatic impacts for the three structural alternatives are discussed below in greater detail. All structural alternatives may negatively impact the economic development potential to the local area. As indicated by Gannett Fleming, Inc, in their permit application submittal: "without addressing the CSO problem, the potential success of the desired water based recreational activities and real economic development will be impaired." To quantify and compare the socio-economic benefits that the inflatable structure will have on the Wyoming Valley Region, in relation to other non-structural alternatives, the Baltimore District engaged the Corps of Engineers to verify the economic benefits anticipated from each alternative.

RED benefits are benefits to a local community or regional economy, such as increases in employment, sales, incomes or sales revenues. Table 7.1 summarizes the Corps estimated RED benefits for three alternatives.

The Corps has estimated RED benefits to the region from the Riverfront Development Plan alternative to be approximately $14.8 million annually, whereas RED benefits to the region from inflatable structure alternative are estimated to be $3.1 million annually.

If water quality issues occur within the impoundment, there will be a corresponding decrease in recreational usage and the expected RED of the applicant's preferred alternative, may not accrue as expected.

The proposed project may also negatively impact the exiting economic benefit to the citizens of New York from the exiting walleye fishery. The existing Susquehanna River's Recreational walleye fishery is one of the best in New York State and supports a modest but important segment of the local economy within the Binghamton Metropolitan Area. An inflatable structure at Wilkes-Barre will impede normal migration patterns of walleye and impact the quality of the fishery in New York State.

Loss of future expected fishing revenue within the Susquehanna River basin will also occur as a result of the proposed project, because it will not provide adequate anadromous fish passage.

Since the RED benefits for the West Bank Riverfront Development Plan component of this alternative were not calculated, the District concludes that the RED benefits will be

Table 7.1: Economics Review of the WBIS Permit Application

Wyoming Valley Inflatable Structure Aternatives	*Expected Visitation Numbers*	*RED Benefits County Tax Revenue*	*RED Benefits County Sales (direct)*	*RED Average Annual Benefits*
Riverfront Development Plan	330,275	$2.8 million	$12.0 million	$14.8 million
Inflatable Structure	69,725	$591,600	$2.5 million	$3.1 million
Riverfront Development Plan + Inflatable Structure	400,000	$3.4 million	$14.5 million	$17.9 million
Riverfront Development Plan + Inflatable Structure + West Bank Riverfront Development Plan	Expected visitation numbers and RED benefits for this alternative are not available because data for West Bank Riverfront Development was unavailable. Benefits would be at least $17.9 million—the amount shown for the Riverfront Development Plan + Inflatable Structure			

grater than $14.8 millions (Riverfront Development Plan) for this alternative, substantially meeting the overall project purpose defined by the Corps. The Corps has determined that this alternative will provide approximately 83% of the RED benefits defined by the applicant's preferred alternative. This alternative will not have the immediate or long term costs associated with construction and maintenance of an inflatable structure and fish passage facility. Significant river-based recreational opportunities will occur with the Riverfront Development Plan and the West Bank Riverfront Development Plan. This alternative meets the overall project purpose as defined by the Corps.

8

The Motts Run Water Treatment Plant[1]

Spotsylvania County

Editor's Introduction: Large storage dams Bhakra and Tehri store water in monsoons, when the supply is large, and make it available during summer, when the demand is high. The same objective of inter-seasonal transfer of water can be attained without damming the river. Water in excess of specified minimum monsoon flow can be diverted from the river and stored in an artificial lake. This artificial lake can be emptied during the summers for producing hydropower as well as for providing water. The main river can flow uninterrupted. Such a side-stream skimmer has been built by the City of Fredericksburg in Virgina. Water from River Rappahannock is pumped into the Motts Run Reservoir during high-flows and then used for supplying water to the city. River Rappahannock flows uninterrupted. This model can help us decommission large

1. Adapted from www.spotsylvania.va.us/emplibrary/2Background.pdf

storage dams like Bhakra, Sardar Sarovar and Tehri and restore free flow of rivers while maintaining supply of water.

The Spotsylvania County water distribution system has seen significant growth in recent years, especially with the installation and operation of the Motts Run Water Treatment Plant (WTP). This plant, owned by Spotsylvania County, serves both Spotsylvania County and the City of Fredericksburg, and significantly altered the operation of both water systems.

In 2000, Spotsylvania County placed the Motts Run WTP into service and the City of Fredericksburg decommissioning the Kenmore WTP located along the Rappahannock River. The City currently receives its primary water supply from the Motts Run WTP through a connection at Falls Hill Avenue and continues to maintain two emergency connections to Stafford County.

Spotsylvania County water system consists of the following principal features:

Ni Reservoir

The Ni Reservoir, the raw water supply to the adjacent Ni Water Treatment Plant, was constructed in 1974 by the U.S. Department of Agriculture, Soil and Conservation Service. The reservoir has a volume of 750 million gallons covering approximately 417 acres. The safe yield, as defined by the Virginia Water Control Board, is 4.0 mgd annual average withdrawal. Safe yield was re-calculated in 1999 during the 1998-1999 drought at 4.3 mgd; however, the 1998-1999 drought was not confirmed to be the drought of record.

The class 2 earth-fill dam at the Ni Reservoir is 45 feet high and 1440 feet long. The dam principle spillway is at an elevation of 237.5 feet, while the emergency spillway is at 247.7 feet.

The Ni WTP was initially constructed in 1974 with a 1.0 mgd capacity. The Ni WTP includes a rapid mix facility, three parallel up flow clarification processes, mixed media filtration, and clearwell storage.

The clearwell storage consists of three separate underground structures containing a total volume of 358,000 gallons. Raw water pumping consists of four 2- mgd raw water pumps, with one used as a backup. The finished water pumping consists of four pumps - at 1300 gpm. An additional 1400-gpm pump is stored and maintained at the plant, to be utilized if one of the installed pumps goes out of service for a prolonged period.

Before the Motts Run Water Treatment Plant was placed into service in 2000, the Ni WTP served as the primary water supply to Spotsylvania County.

Motts Run Reservoir

The Motts Run Reservoir was built in 1971. The reservoir is located adjacent to the Rappahannock River; prior to 1995 it was owned and maintained solely by the City of Fredericksburg. Currently, the reservoir is owned by the City of Fredericksburg, but is jointly operated by the City of Fredericksburg and Spotsylvania County. The total reservoir volume, prior to modifications, is 1.3 billion gallons.

The existing dam is of earth-fill construction is approximately 100 feet high and 700 feet long. Recent modifications to the dam include replacement of two 10-inch intake tower gates with two 24-inch sluice gates, slip-lining the existing 48-inch conduit with 700 feet of 36-inch polyethylene pipe, upgrading the emergency spillway, reshaping the dam, armoring the dam side to reduce erosion, raising the crest of the dam 3.5 feet, and installing a blanket drain system on the downstream side. Upon completion of the dam modifications, the reservoir will be capable of supplying raw water to the Motts Run WTP or receiving water to be stored from the Rappahannock River raw water pump station and intake. Safe yield of this reservoir varies based on natural inflow assumptions, inflow pumping rates, and sedimentation and recreational storage requirements. Applying conservative criteria and assumptions the safe yield has been estimated to be between 3.5 and 4.0 mgd.

The Motts Run water treatment facility and Rappahannock River raw water pumping station were completed in the spring of 2000. The current water treatment plant capacity is 12 mgd and is expandable to 24 mgd.

The Rappahannock River Intake facilities consist of a 36-inch diameter raw water conduit from a header system of three (expandable to 4) cylindrical shoreline well screens to the raw water pumping station. The riverbank inlet includes both de-icing and automatic air bursting systems to maintain consistent intake flows. The raw water pumping station, located adjacent to the intake has a Rappahannock River withdrawal capacity of 28 mgd. Raw water can either be pumped to the Motts Run WTP via vertical turbine pumps or to the Motts Run Reservoir through two 10-mgd vertical turbine pumps. As an alternative, water can also be released from the Motts Run Reservoir into the Rappahannock River upstream of the intake structure or can be piped directly to the raw water pumping station to be treated at the plant.

PART-III

COST-BENEFIT ANALYSIS OF HYDROPOWER DAMS

9

Cost-Benefit Analysis[1]

Office of Management and Budget

Editor's Introduction: This paper is extracted from a rather long U.S. Government guideline for undertaking Cost-Benefit Analysis of regulatory actions. The original paper was written to explain how to assess whether a particular regulation, such as requiring a particular type of refrigerator, is desirable. The same CBA method is applicable for dams. The learnings from this paper are given below.

Make a schedule of non-quantitative costs and benefits. Our projects ignore non-quantitative costs and benefits. For example, the loss of soul due to submergence of a temple is not internalized. Or, the loss of mental peace from deterioration of quality of river water due to passing through turbines is not internalized in our CBAs. This paper says a separate schedule of non-quantitative costs and benefits must be attached to make it possible for policy makers to appreciate these aspects.

---

1. Adapted from Circular A-4 dated September 17, 2003 issued by office of Management and Budget, U.S. Government.

Specify if gains and losses occur to different groups. Dams impose costs such as those of increased incidence of malaria on people near the reservoirs while they provide benefits such as electricity or irrigation to user groups. The Tehri Dam, for example, imposes costs on villages around the reservoir rim while it provides drinking water to Delhi. This must be explicitly mentioned. It is not enough to state that total benefits are greater than total costs. It is possible that one section of the people have to bear net costs while another section gets net benefits.

Specify if intergenerational equity is involved. Dams have finite life even though some may last for 100 years or more. At some point future generations will inherit non-functioning dams along with broken rivers. The present generations will benefit at the cost of these future generations. This tradeoff should be explicitly stated.

Various alternatives must be examined. It is customary for our regulatory agencies to provide a single CBA. It is shown that costs are less and benefits are more for a given project. The Office of Management and Budget (OMB) states such analysis is inadequate. It is necessary to examine various alternatives. It is necessary to compare benefits and costs with other alternatives such as demand side management, solar power, micro hydel etc. This is necessary so that second-best projects are not implemented. A particular hydropower project may have Cost-Benefit Ratio of, say, 2.0. It should not be implemented on this basis. It is possible that Cost-Benefit Ratio of demand side management may be 2.5. In that case, demand side management must be implemented instead of making a hydropower project.

Opportunity values must be specified. The resources used in making dams—rivers, forests and wildlife would provide some benefits to the people if dams were not made. This concept is accepted by Government of India in requiring users of forests to deposit 'Net Present Value' of services provided by forests with the Government. This concept should be applied to all resources used such as rivers and wildlife.

The Central Electricity Authority states that there is shortage of power hence hydropower dams should be built. It does not specify what is the price society will have to pay if power is not produced from a particular hydropower dam. This benefit will be the difference between (1) the price at which power is produced from a project; and (2) the price at which power would be available from next cheapest alternative source. A hydropower project may provide power at, say, Rs 3.00 per KWH. The next cheapest source, say, solar power, provides the same power at Rs 5.00 per unit. The benefit from the project would then be Rs 2.00 per unit. This must be quantified and it should then be examined whether the costs of the project are justified. It is insufficient to implement a project merely because there is shortage of power.

Uncertainty must be specified. There are elements of uncertainty in every project. For example, development of solar power at Rs 4.00 per KWH, as proposed by some non-profit players, would render many hydropower projects uneconomic. Or development of LED bulbs can reduce peak demand. The flow in our rivers may reduce due to recession of glaciers. A statement of such uncertainties and how they will affect the project should be provided.

This Circular is designed to assist in defining good regulatory analysis and standardizing the way benefits and costs are measured and reported.

Benefit-Cost Analysis

A distinctive feature of BCA is that both benefits and costs are expressed in monetary units which allow you to evaluate different regulatory options with a variety of attributes using a common measure. By measuring incremental benefits and costs of successively more stringent regulatory alternatives, you can identify the alternative that maximizes net benefits.

The size of net benefits, the absolute difference between the projected benefits and costs, indicates whether one policy is more efficient than another. The ratio of benefits to costs is not a meaningful indicator of net benefits and should not be used for that purpose. It is well known that considering such ratios alone can yield misleading results.

You should exercise professional judgment in identifying the importance of non-quantified factors and assess as best you can how they might change the ranking of alternatives based on estimated net benefits. If the non-quantified benefits and costs are likely to be important, you should recommend which of the non-quantified factors are of sufficient importance to justify consideration in the regulatory decision. This discussion should also include a clear explanation that support designating these non-quantified factors as important. In this case, you should also consider conducting a threshold analysis to help decision makers and other users of the analysis to understand the potential significance of these factors to the overall analysis.

Distributional Effects

Those who bear the costs of a regulation and those who enjoy its benefits often are not the same people. The term "distributional effect" refers to the impact of a regulatory action across the population and economy, divided up in various ways (e.g., income groups, race, sex, industrial sector, geography). Benefits and costs of a regulation may also be distributed unevenly over time, perhaps spanning several generations.

Your regulatory analysis should provide a separate description of distributional effects (i.e., how both benefits and costs are distributed among sub-populations of particular concern) so that decision makers can properly consider them along with the effects on economic efficiency. Executive Order 12866 authorizes this approach. Where distributive effects are thought to be important, the effects of various regulatory alternatives should be described quantitatively to the extent possible, including the magnitude, likelihood, and severity of impacts on particular groups. You should be alert for situations in which regulatory alternatives result in significant changes in treatment or outcomes for different groups. Your analysis should also present information on the streams of benefits and costs over time in order to provide a basis for assessing intertemporal distributional consequences, particularly where intergenerational effects are concerned.

Evaluation of Alternatives

You should describe the alternatives available to you and the reasons for choosing one alternative over another. As noted previously, alternatives that rely on incentives and offer increased flexibility are often more cost-effective than more prescriptive approaches. For instance, user fees and information dissemination may be good alternatives to direct command-and-control regulation. Within a command-and-control regulatory program, performance-based standards generally offer advantages over standards specifying design, behavior, or manner of compliance.

You should carefully consider all appropriate alternatives for the key attributes or provisions of the rule. It is not adequate simply to report a comparison of the agency's preferred option to the chosen baseline. Whenever you report the benefits and costs of alternative options, you should present both total and incremental benefits and costs. You should present incremental benefits and costs as differences from the corresponding estimates associated with the next less-stringent alternative. It is important to emphasize that incremental effects are simply differences between successively more stringent alternatives. Results involving a comparison to a "next best" alternative may be especially useful.

In some cases, you may decide to analyze a wide array of options. In 1998, DOE analyzed a large number of options in setting new energy efficiency standards for refrigerators and freezers and produced a rich amount of information on their relative effects. This analysis - examining more than 20 alternative performance standards for one class of refrigerators with top-mounted freezers — enabled DOE to select an option that produced $200 more in estimated net benefits per refrigerator than the least attractive option.

Developing Benefit and Cost Estimates

The analysis document should discuss the expected benefits and costs of the selected regulatory option and any reasonable alternatives. To present your results, you should:

- Include separate schedules of the monetized benefits and costs that show the type and timing of benefits and costs, and express the estimates in this table in constant, undiscounted dollars.
- List the benefits and costs you can quantify, but cannot monetize, including their timing; describe benefits and costs you cannot quantify; and
- Identify or cross-reference the data or studies on which you base the benefit and cost estimates.

When benefit and cost estimates are uncertain you should report benefit and cost estimates that reflect the full probability distribution of potential consequences.

If fundamental scientific disagreement or lack of knowledge prevents construction of a scientifically defensible probability distribution, you should describe benefits or costs under plausible scenarios and characterize the evidence and assumptions underlying each alternative scenano.

Key Concepts

"Opportunity cost" is the appropriate concept for valuing both benefits and costs. The principle of "willingness-to-pay" (WTP) captures the notion of opportunity cost by measuring what individuals are willing to forgo to enjoy a particular benefit. In general, economists tend to view WTP as the most appropriate measure of opportunity cost, but an individual's "willingness-to-accept" (WTA) compensation for not receiving the improvement can also provide a valid measure of opportunity cost.

The use of any resource has an opportunity cost regardless of whether the resource is already owned or has to be purchased. That opportunity cost is equal to the net benefit the resource would have provided in the absence of the requirement. For example, if regulation of an industrial plant affects the use of additional land or buildings within the existing plant boundary, the cost analysis should include the opportunity cost of using the additional land or facilities.

To the extent possible, you should monetize any such forgone benefits and add them to the other costs of that alternative.

Estimating benefits and costs when market prices are hard to measure or markets do not exist is more difficult. In these cases, you need to develop appropriate proxies that simulate market exchange. Estimates of willingness-to-pay based on revealed preference methods can be quite useful.

Methods for Treating Non-Monetized Benefits and Costs

Sound quantitative estimates of benefits and costs, where feasible, are preferable to qualitative descriptions of benefits and costs because they help decision makers understand the magnitudes of the effects of alternative actions. However, some important benefits and costs (e.g., privacy protection) may be inherently too difficult to quantify or monetize given current data and methods. You should carry out a careful evaluation of non-quantified benefits and costs. Some authorities refer to these non-monetized and non-quantified effects as "intangible".

You should monetize quantitative estimates whenever possible. Use sound and defensible values or procedures to monetize benefits and costs, and ensure that key analytical assumptions are defensible. If monetization is impossible, explain why and present all available quantitative information. For example, if you can quantify but cannot monetize increases in water quality and fish populations resulting from water quality regulation, you can describe benefits in terms of stream miles of improved water quality for boaters and increases in game fish populations for anglers.

If you are not able to quantify the effects, you should present any relevant quantitative information along with a description of the unquantified effects, such as ecological gains, improvements in quality of life, and aesthetic beauty. You should provide a discussion of the strengths and limitations of the qualitative information. This should include information on the key reason(s) why they cannot be quantified. In one instance, you may know with certainty the magnitude of a risk to which a substantial, but unknown, number of individuals are exposed. In another instance, the existence of a risk may be based on highly speculative assumptions, and the magnitude of the risk may be unknown.

For cases in which the unquantified benefits or costs affect a policy choice, you should provide a clear explanation of the rationale behind the choice. Such an explanation could include detailed information on the nature, timing, likelihood, location, and distribution of the unquantified benefits and costs. Also, please include a summary table that lists all the unquantified benefits and costs, and use your professional judgment to highlight those that you believe are most important.

10

Economic Value of Hydropower[1]

David Marcus

Editor's Introduction: This is a critique of FERC's approvals of hydropower projects. The contents of this paper will be appreciated if the learned reader keeps three levels of discourse in mind.

1. The practice by CEA or MOEF in India while approving hydropower projects. This is the least acceptable situation.
2. The practice followed by FERC in the U.S. This is one step better than India. FERC at least makes a notional effort to examine the benefits of power produced and costs of environmental mitigation measures proposed.
3. The critique of FERC practice by the author of present paper.

The author's critique indicates how far we have to go. We have to first start assessing the various costs and benefits of

1. Adapted from "FERC's economic analysis of hydro projects : A review of policy and practice since the Mead decision," March 18, 1997.

the projects as being done by FERC presently. Next we have to do this in a true manner as suggested by this author.

The author says that FERC rejects proposed mitigation measures if they are expensive without examining the benefits. In other words, the sole objective is to ensure commercial feasibility of the project. Environmental mitigation measures that would possibly lead to cancellation of projects are rejected simply because they are 'too expensive'. CEA and MOEF do not undertake such analysis at all.

The social benefit from production of power from a particular project is the difference between the cost of production of power from the proposed project and the 'replacement cost' at which the same power is available from the next best source. This involves assessment of the replacement cost of power. The author criticizes the FERC for overestimating the replacement cost and thus approving projects for which cheaper alternate sources are available. In India the CEA does not even make such an analysis. Mere existence of 'shortage' is considered adequate reason for approving projects irrespective of whether the costs are much higher than benefit. CEA does not undertake a cost-benefit analysis of projects it approves.

The FERC assumes that cheap power produced by a proposed project will be wholly replaced by expensive power from next best alternative source. Let us say a project is designed to produce 1 million units of power at Rs 2 per unit. In absence of the proposed project, this 1 million units would be produced by alternative sources, say at Rs 3 per unit. The benefit from the project is thus estimated to be Rs 1 million (at Rs 1 per unit). The author argues this calculation is not correct. The demand for power at Rs 3 per unit will be less than 1 million units hence the replacement cost will be less. Our CEA similarly merely states 'x' amount of shortage of power exists. It does not examine how the shortage will behave if the true price of power is charged from the consumer.

A market for electricity is already in operation in India. According to one news report the average cost of power that was traded was Rs 4.50 per unit in 2007-08. This is a good estimate of replacement price of power.

Other costs not included by both—FERC and CEA—include the following:

- The social costs and benefits are ignored. Projects that are beneficial for the promoters but possibly harmful for the society are cleared.
- Costs of removal of hydro dams are not internalized.
- Social costs such as those of air pollution are acknowledged but not included in cost calculations.
- Mitigation measures for environmental impacts are prescribed only to such extent that does not jeopardize the project. In this way cost is reduced and project is forcibly made profitable for the promoters.

These comments apply wholly to the clearances given by our Central Electricity Authority.

The Central Electricity Authority does not look into the environmental costs at all. It is left for the Ministry of Environment and Forests to examine the environmental impacts. The economic- and environmental examinations follow two different routes with no interaction whatsoever. This is fallacious because the environmental mitigation measures impact the economic viability of the project. The key lesson for India is that a comprehensive combined clearance authority for economic viability and environment impacts is required.

Summary of findings and recommendations

My review of NEPA documentation and licenses shows that FERC analyzes the economic costs, but rarely the values, of the nonpower aspects of projects. For power generation, both cost and value are calculated. However, power value analyses are frequently flawed by inaccurate assumptions about the source, characteristics, and costs of replacement generation in the event of license denial. Decommissioning costs are rarely taken into account. Cost inflation is (intentionally) ignored, yet future costs are discounted at high rates which make no sense in the absence of inflation. FERC's economic analysis methods facilitate the inappropriate rejection of mitigation measures with large environmental, recreational, or other nonpower benefits. They also make license denial almost impossible, even when FERC openly admits that dam removal would be environmentally preferable.

I conclude that there are several ways FERC could improve its economic analysis. These include a greater use of market prices to evaluate replacement energy costs and use of a lower discount rate for future cost. Decommissioning costs should be incorporated into both FERC's economic calculations and its licensing decisions. Perhaps most importantly, FERC should perform far more quantification of the dollar benefits of the nonpower aspects of projects. Particularly where mitigation measures with both high dollar costs and large environmental benefits are involved, the current subjective balancing of dollars against environmental quality can be greatly improved. FERC should focus on the forward-looking costs and benefits of projects, not their past costs, if only to avoid exacerbating stranded cost problems during this era of electric power restructuring. Finally, FERC should reopen the possibility of letting overall project value affect licensing decisions. FERC should consider whether a project license should ever be denied or further conditioned in response to a strongly negative net project economic value.

Economic analysis of power generation

FERC generally analyzes overall project power economics from a licensee point of view, using total net cost as the measure. Project impacts which do not accrue to the licensee (e.g., air quality benefits) are not included in the overall project economic analysis even when they have been calculated in dollar terms. This method fails to produce useful data from either a licensee or public point of view.

From a licensee point of view, the relative economic comparison is between the forward-looking costs of hydro operation and those of alternative power sources. Sunk capital costs just don't matter. That is why, as FERC finds itself repeatedly forced to acknowledge, licensees accept licenses which FERC has found to have negative economic value. Indeed, licensees themselves are proposing license conditions which either make economic projects become uneconomic (according to FERC) or make uneconomic projects more so.

From the public's point of view, forward looking costs and benefits are also the only ones that matter, but it is both licensee and public-costs and benefits that matter. If imposing a given license condition on the applicant will produce a public benefit equal to or greater than the cost of that condition, then from the public's point of view FERC ought to do so. Only if the private cost to the applicant would be so large as to make the applicant abandon the license entirely (so that the public benefit would in any case not be attainable) should the license condition not be imposed.

How FERC develops its power figures

FERC relies heavily on its staff. In virtually every license decision, the economic figures which appear are taken from staff-prepared environmental documents. Dollar values in final documents usually differ from those in the draft documents only because of changes in the mix of mitigation measures. Only occasionally, where market conditions have changed substantially, do the economic figures change in any basic way.

The basic FERC methodology is to take the unamortized capital cost of the project and some financial assumptions (tax rate, interest rate, discount rate), and compute a levelized annual project capital cost over a 30-year period. To this are added annual O&M costs at current levels and the dollar costs of any proposed license conditions. The resulting total project cost is then subtracted from the cost of replacement power to produce the net annual economic benefit of the project.

The key policy-driven assumptions in FERC's analysis are that no inflation will be included in the calculations, and that all generation would be replaced (i.e., no demand-side impact due to replacement power costs above project costs). The first of these assumptions is well-publicized. FERC has justified dropping all inflation assumptions by pointing out, correctly, that it and everyone else has a terrible track record at trying to predict future inflation in general, let alone future inflation of specific components of power costs such as fuel.

The assumption that 100 percent of a low-cost hydro project's generation would have to be replaced with higher-cost power is reasonable for projects which make up only a small portion of the owner's portfolio. For projects owned by industrial facilities, however, it is likely that the owner would not replace, for example, 34 mill hydro power with 74 mill purchased power on a kwh-for-kwh basis. Surely there are demand-reducing measures with intermediate costs.

Other basic assumptions include the interest and discount rates used to levelize costs (typically 10 percent per year), the project life (30 years), tax assumptions used to compute financing costs, and the replacement technology and its capital cost, financing and heat rate (used for projects assumed to be replaced with new generation).

Where applicants are part of regional pools or are closely interconnected with large regional suppliers with well-known selling prices, FERC has begun to use those prices. For instance, in licensing the Enloe Dam project in Washington FERC used BPA new resource costs (of 28 mills/kwh) as the alternative power cost, superseding much higher figures.

With implementation of FERC Order 888 and state-by-state restructuring, market prices for electricity will be available nationwide. Use of market prices should supersede use of generation proxy costs, since the latter are merely FERC's guess at which of a variety of alternatives would detennine the cost of replacement power.

FERC assumes that replacement power costs would include the cost of replacing 100 percent of both the energy and capacity of the project in question. For an applicant with excess capacity, license denial would require replacement energy but not immediate replacement capacity. In the U.S. today, particularly in the West, many utilities have excess capacity. Thus, the appropriate cost to use for alternative power cost analyses should be the current cost of energy plus the cost of capacity **bought at some time in the future,** and appropriately discounted back to the present.

Failure to perform analysis

There is a well-established literature on the nonpower costs and benefits of dams. HRC has specifically requested that I not focus in this paper on how to do nonpower economic analysis, but rather on how FERC incorporates nonpower values into its analysis. The answer is simple: FERC does virtually no quantitative analysis of nonpower values in dollar terms, yet routinely "balances" the dollar costs of nonpower mitigation measures against their environmental benefits.

How can FERC "balance" without analysis? Usually just by assertion. Measures are repeatedly adopted or rejected in environmental documents based on staff's judgment that their economic costs outweigh their non-economic benefits, without any attempt to quantify their non-economic benefits in dollar terms. In other cases, mitigation measures are rejected because they have "high" costs, without any analysis of whether they may also have high benefits. In yet other cases, benefits are asserted to be low or hard to measure, and are rejected on that basis, even when there is quantitative analysis from intervenors which does express benefits in dollar terms.

Finally, some measures are rejected based on cost relative to net overall project benefits. For example, an October 1996 FERC license for the Penobscot Mills project finds that the project as licensed will have a net positive benefit to the licensee of 39.62 mill/kwh, yet rejects a measure to create salmon habitat which would cost 2.44 mill/kwh on the grounds that "the modest fisheries benefit likely to occur" does not justify "the **significant** adverse impact on the project's energy benefits". But FERC makes no effort whatsoever to explain why a 2.44 mill/kwh (6 percent) benefit reduction would be "significant" for the project with the greatest net benefits of any of the 18 licenses I have reviewed.

With the exception of air quality, the FERC documents I have reviewed have virtually no quantification of any nonpower value in dollar terms. I describe some exceptions below. In contrast, the three intervenor documents I reviewed each quantified nonpower values which were larger than the cost of achieving them.

Air pollution values are by far the most frequently quantified nonpower value. In many cases, FERC calculates the tons of certain air pollutants whose emission is avoided by the project in question, and then calculates the dollar cost of preventing the emission of that many tons of pollutant. The result is a dollar valuation of the benefit of the project in terms of air pollution avoidance. While there are various theoretical objections to FERC's methodology, it is certainly a reasonable approach.

The trouble is, the resultant dollar values are not used for anything. They are not incorporated into the calculation of net project benefits, because FERC does that on an applicant basis rather than a societal basis. They are not incorporated into a balancing calculation, because there are no other environmental consequences of the project which have been calculated on dollar terms. They are just calculated and printed.

In individual proceedings FERC has quantified in dollar terms the nonpower values associated with fish entrainment (and its avoidance), fish passage benefits, cabin rentals, and recreation benefits. Fish entrainment values, where calculated, are used to provide a quantitative basis for rejection of fish screens or other physical measures to prevent fish entrainment.

Use of non power economic values in FERC analysis

Measures intended to increase nonpower values are routinely included in licenses as conditions, but the basis for doing so appears to be qualitative and ad hoc. Just to give one example, the DEIS for four projects on the Menominee River includes a package of mitigation measures which "would enhance fisheries and recreation values," while conceding "the effects cannot be quantified, but we do not expect them to be significant..."

FERC's reluctance to quantify the dollar value of mitigation measures does not extend to quantifying their costs. For each measure included in a particular alternative, FERC tries to quantify the dollar cost of implementing that measure. Thus, the dollar costs of the staff-proposed mitigation measures are included in the economic analysis of the staff proposal.

Consequences of not quantifying nonpower values in dollar terms

By not estimating the economic value of nonpower attributes of the projects it reviews, FERC incorrectly rejects high-cost mitigation measures which would also have large nonpower benefits. At the same time, it may be accepting "routine" low-cost "mitigation" measures whose nonpower benefits are even lower than their cost. In reviewing measures proposed by agencies under FPA sections FERC rarely rejects a low-cost measure.

Most importantly, the absence of nonpower economic analysis means that FERC has abdicated to its staffs intuition and prejudices the balancing of economics and the environment which it admits is its duty under section 10(a). By simply writing in an EA or EIS that a given measure is too expensive relative to its benefits staff gives the Commission the ability to avoid imposing controversial mitigation conditions without any quantitative basis for doing so.

Decommissioning analysis

FERC has made clear that it will not normally order decommissioning or decommissioning funding. In particular cases, it has even refused to order **studies** of the cost of decommissioning. Thus, the decommissioning analysis in almost all FERC proceedings is minimal at best, usually just a rejection of the decommissioning alternative with a citation to the Commission's policy statement of 12/14/94.

At the same time, FERC has made clear that, for relicensings, the alternative to license approval is decommissioning, that decommissioning can be expensive, and that the expense of decommissioning may cause licensees to accept a license with conditions which would be economically untenable in the case of a new license.

Without knowing the cost of decommissioning, FERC cannot actually identify whether either the public or the applicant will gain economically from the granting of a license.

The reason is simple. In the absence of a license, decommissioning will have to occur. Licensing therefore saves the cost of alternative sources of power, but it also saves the cost of decommissioning. That is one reason, as FERC has acknowledged, that applicants can be willing to accept licenses whose FERC-computed net benefits are negative. Applicants know that they are saving decommissioning costs, so that an apparently negative-benefit license is still a better deal for them than no license at all, and the associated decommissioning costs.

How FERC uses its economic analysis

FERC staff usually summarizes its project-level economic analysis by providing data on the cost of alternative power and the net cost of the overall project under three or four scenarios. These scenarios consist of relicensing with no additional mitigation (no-action) and relicensing with applicant-proposed mitigation, staff-proposed mitigation, and, where applicable, intervenor- or agency-proposed mitigation. Cost data are provided in terms of either dollars and/or mills/kwh.

No use is made of the figures which result from the project-level economic analysis. In 40 percent of the cases I have examined, FERC staff finds the project to have a net negative value with either the applicant-proposed or the staff-proposed mitigation.

There is a clear policy bias toward issuing a license in all cases. FERC is clearly willing to condition licenses, perhaps even to the point that an applicant will reject one, but has consistently shied away from taking the position that there is any project whose impacts cannot be mitigated.

On individual mitigation measures, there is no clear policy. De facto policy appears to be to start from the Applicant's position, add agency mitigation pursuant to FPA sections 4(e), 10(a), 10(j) and 18, and then make adjustments from there. FERC is clearly willing to reject agency-proposed mitigation based on cost, and routinely rejects or ignores intervenor-proposed mitigation measures.

Changes to present analytical practices

What can FERC do, short of revisiting its policy statement on decommissioning? One simple analytic measure would be to include the cost of decommissioning in its economic analyses even if it is not going to be included as a license condition. Analysts may argue that the cost of decommissioning 30 years in the future will be low in current terms; that's fine, but it's not zero.

Analyzing decommissioning without requiring decommissioning funding would be consistent with FERC's policy statement on decommissioning. FERC routinely analyzes other mitigation measures which it does not adopt, and it routinely analyzes the cost of replacement power which the Applicant will not actually buy if the license is granted. Analyzing decommissioning would allow FERC to find those few cases where license denial is appropriate.

FPA section 10(a) requires FERC to balance different values, power and non-power, in evaluating whether to issue a license. The same balancing is applied to individual mitigation measures. This balancing is currently occurring, in principle. FERC routinely says a given measure is included because its environmental benefits outweigh its costs, or is rejected because its benefits do not outweigh its costs. But there's normally no quantitative analysis. When there is, it sometimes consists of measuring the economic cost of the measure against the economic benefits of the project instead of against the environmental benefits of the measure.

By expressing at least some non-power values in dollar terms, as it already does for the power output of projects, FERC would be able to do less apples-to-oranges comparisons under its balancing mandate, and more apples-to-apples comparisons. This would not be a change in policy, but only in analytical methodology. FERC already converts power values from their natural energy and capacity units (Mw and kwh) into dollar terms; it would just be doing the same thing for fish, recreation, and other non-power products which dams produce or could produce.

FERC may argue that nonpower values are not easily measured because "no direct market exists for most nondevelopmental resources." "Most" is not "all." There are river-based resources other than power that are market-priced resources, commercial fisheries and commercial rafting for example. Others have well-developed methodologies for computation of their economic value, such as recreational benefit analysis.

Where standard methods exist, and where the nonpower resource is itself a commercial resource with prices, there is no reason not to convert nonpower values (e.g., user-days of whitewater rafting, pounds of fish caught by commercial fishermen) into dollar terms. In particular, when another party has already done the analysis, there is really no excuse for FERC not to provide dollar values.

Policy changes

There are several actions the Commission can take to reflect the real possibility of decommissioning, whether sooner or later, which will not require changing the existing policy statement.

FERC should start including future decommissioning costs as part of the cost of a project in evaluating project economics. Relicensings are not forever. Even if project owners are excused from pre-funding decommissioning on the grounds that their financial solvency is not in doubt, they will still have to pay decommissioning costs at the end of their license term.

The fact that decommissioning costs 30 years in the future will be greatly minimized by the discounting process is true, but is no reason to ignore them, particularly if discount rates of less than 10 percent are going to be used.

FERC should do its analysis of project economics with near-term decommissioning as part of the cost of alternative power sources. In the same way that project benefits include the dollars saved by not having to acquire alternative sources of power, project benefits also include the dollars saved by deferring for 30 or 40 years the cost of decommissioning.

Where it has done a serious analysis of decommissioning as an alternative, FERC has already done precisely the kind of analysis suggested in the preceding paragraph. The trouble is that FERC has rarely considered decommissioning seriously. Even in cases where decommissioning is not going to be adopted, including decommissioning costs in the economic analysis would give a much more accurate view of the project's economics. At the very least, decommissioning costs should be analyzed in cases where project economics are already negative without decommissioning.

FERC has made clear its intention to always try to craft conditions such that a proposed project will comply with FPA section 10(a). If the consequences of doing so result in a project whose economic costs are greater than those associated with decommissioning, and whose environmental benefits are no greater than those associated with decommissioning, then the Commission should choose decommissioning.

While it may seem obvious that FERC should not license a project which is both economically and environmentally inferior than no project at all, it does not seem to be current policy. Rather, FERC's position seems to be that if it can find mitigation measures which offset the negative impacts that a project would otherwise have, then it will offer a license no matter how much those measures cost. It will thus become the applicant's choice whether mitigation is more costly than decommissioning, a choice that FERC itself ought to be making.

FERC staff defends its routine failure to quantify the benefits of mitigation measures in dollar terms by saying that such analysis is difficult to do and subject to controversy. So is power value analysis. Even if FERC thinks quantifying the dollar benefits of mitigation measures is generally beyond its capabilities, it should do such analysis as a matter of policy when major mitigation options offer large benefits in exchange for large costs.

In a relatively small number of FERC proceedings, major changes in the project are proposed by intervenors or agencies which are intended to have equally major environmental benefits. Dam dismantlement is the most extreme example of such major changes. In such cases, where the economic costs are high but the environmental benefit is also high, FERC should quantitatively analyze the dollar value of the environmental benefit.

Section l0(a) of the FPA requires FERC to balance nonpower values against power values. If power values are converted into dollars, but nonpower values are not, then the section 10(a) balancing is necessarily a comparison of apples and oranges. Such a comparison is certainly legal, but it is difficult and open to the perception of arbitrariness and capriciousness. It will be necessary in many if not all cases, since many impacts of hydro projects cannot be reduced to dollar terms. But if the tools for comparison on a dollar-to-dollar basis exist, FERC should use them, at least when the stakes are large.

FERC's position is that it will not deny a project based on apparently negative economics, but will leave to the applicant the decision as to whether to accept the license as offered. Given that decision, why should FERC even bother analyzing the economics of applications from an applicant point of view? Since the stated purpose of economic analysis is now to aid in the 10(a) balancing analysis, economic analysis should be done from a public point of view. That probably would imply lower discount rates and, most importantly, quantification of environmental and recreational benefits to the public which would result from alternative licensing conditions or decisions. An analysis which computes the costs of measures to the applicant but not their benefits to the public does not adequately portray the impact of the project from the public's point of view.

11

Critique of FERC's Approval to Klamath Hydroelectric Dam[1]

Jon H. Goldstein

Editor's Introduction: This paper gives a critique of the decision of FERC, which is equivalent to our CEA, to relicense the Klamath Hydroelectric Project. On some issues, the FERC is ahead of our CEA. On others the FERC and CEA are in the same boat.

Issues on which the U.S. Regulators are ahead of CEA

The Oregon PUC rejected PacifiCorp's request for establishing a thermal power plant saying that PacifiCorp had overstated its need for power and should pursue short-term market purchases, renewable energy, demand-side management and efficiency measures. Our CEA gives lip service

1 Adapted from "Assessment of Deficiencies, Omissions, and Distortions in the Economic and Power Sections of FERC DEIS for the Relicensing of the Klamath Hydroelectric Project: An Annotated Guide to Source Material," American Rivers, June 5, 2007.

to measures such as reduction of T&D losses but does not see them as alternative to production of more power.

When FERC reviews licensing proposals, in addition to assessing power and other developmental uses of the waterway (such as irrigation and water supply), it gives consideration to energy conservation, fish and wildlife habitat, recreational opportunities, and environmental quality. These are integral components of project appraisal. In India, however, these aspects are not examined by CEA and left to MOEF. Hence a comprehensive assessment of benefits from power and costs to environment is not made.

FERC considers the decommissioning and sediment removal costs. CEA and MOEF do not consider these.

FERC gives findings on objections raised during licensing proceedings. These are like a 'speaking order'. The CEA has no window for public participation. Even the Expert Appraisal Committee of MOEF gives no speaking orders. It only notes that 'comments made during public hearings have been considered.' Thus public is denied knowledge of reasons for not accepting the objections raised by them.

FERC does make an analysis of replacement cost for power even though the figures may be somewhat manipulated. CEA, on the other hand, considers existence of 'shortage' as a self-evident reason for producing more power. It does not examine the cost to the economy from the shortage and the cost to economy from production of more power.

Issues on which U.S. Regulators and CEA are in the same boat.

Both do not examine the impact of environmental mitigation measures on the production of power. For example, minimum flow requirements reduce the availability of power and increase its price. These are not assessed and projects are cleared assuming no minimum flow requirements are required.

The requirement of power varies between seasons. The projects are cleared without examining this aspect. In India, hydropower projects produce most power during monsoons when the demand is less and, therefore, the benefit to consumers is less. 'Shortage' assessed by CEA may relate to non-monsoon period while the project cleared would produce most power during monsoon period.

High price of short-term or spot market purchase of power are invoked as justification for making hydropower projects. The Chief Minister of one state told this author that he needed to build more hydropower projects because he had to buy power at rates as high as Rs. 8.50 per unit. But it is not examined whether the projects established would produce power when precisely these circumstances are obtained. In other words, the spot price of power is used as justification to produce more base power.

Production of power by hydro projects depends upon the releases from upstream dams and requirements of downstream water users. These restrictions are not internalized by CEA while approving projects. For example, CEA approved the Kotli Bhel projects on Alaknanda assuming a continuous river flow and ignored the fact that upstream Srinagar project would release waters only during peak hours.

Under the Federal Power Act (FPA) as amended by ECPA, FERC is charged with making decisions about hydropower projects that are in the public interest, including giving equal consideration to power and non-power benefits. Despite this legislative mandate, in preparing EISs FERC consistently:

- Selects replacement cost for power in excess of power value (e.g., in excess of avoided cost);
- Avoids quantifying non-power benefits as well as resource losses and damages;
- Asserts that the benefits derived from mitigation proposed by stakeholders are not worth the costs of conducting the mitigation. FERC does this without disclosing the metrics used in assigning value to these non-power benefits; and
- Refuses as a matter of policy to provide documentation for its calculations, assertions, and conclusions, even to

other Government agencies with regulatory responsibilities in the same proceeding.

Structured thus, FERC's EIS process biases the analytic results and decisions in favor of relicensing a hydroelectric project with limited mitigation and restrictions. Such a process supports power production over other, non-power uses (market and non-market) of the project and waterway, and does not result in decisions that are in the public interest.

FERC's DEIS for the relicensing of the Klamath Hydroelectric Project is the quintessential example of these actions. In particular, the DEIS:

- Exaggerates the need for power in the region served by the Klamath Project and the ability of a relicensed Klamath to contribute to such need.
- Exaggerates the dependability of the Project to provide power on demand. The DEIS does this despite evidence to the contrary by PacifiCorp in both the Final License Application and other forums.
- Exaggerates the proportion of the Project's power which is generated during peak hours. Indeed, FERC increased without explanation the proportion claimed by PacifiCorp, a proportion which was already suspect.
- Exaggerates the cost of replacing Klamath's power, a critical variable in deciding on mitigation measures and relicensing conditions.
- Ignores efficiency upgrades as a procedure for offsetting reduced generation from the Klamath.
- Superficially assesses the decommissioning option despite being forewarned by stakeholders early in the licensing process that decommissioning should be given serious consideration.

In addition to the deficiencies in the DEIS, FERC:

- Refused requests for documentation of the analysis in its DEIS, an apparent violation of its core responsibilities under NEPA.

- Failed to acknowledge, let alone systematically consider, cheaper, more efficient alternatives for replacing Klamath's power, a violation of its responsibilities under Section 10(j) of the FPA.

Focusing on the power and economic aspects of the Klamath Project, this paper documents the key flaws in the Klamath DEIS with respect to power generation, power values, and the economics of relicensing the project with mitigation versus decommissioning it. The paper serves not only as a guide to the DEIS but also as a guide to critical assessments by state and federal resource agencies and other stakeholders.

Klamath as a Source of Peak Power

As noted above, although Klamath has some peaking capacity, it is primarily a run-of-the-river facility. It is a low-head hydro facility with shallow reservoirs and minimal storage capacity. Thus, as a provider of peak power it is largely an opportunistic facility with some limited ability to store water and manage its reservoirs and generation.

In the Final License Application, PacifiCorp estimated that 64% of Klamath's generation is on-peak. The CEC found PacifiCorp's division of generation into peak and off-peak too imprecise, arguing that the day should be more finely divided (super peak, shoulder 1, shoulder 2, off-peak), because the differences in prices at various times is significant.

Consequences if peak generation is exaggerated: If too much of a project's generation is classified as peak, its annual power value will be inflated, because in calculating power value too much of the power will be priced at peak rates.

FERC's estimate of Klamath's generation during peak hours: FERC did not share the CEC's concerns about PacifiCorp's identification of Klamath's generation as 64% peak power. FERC did not ask PacifiCorp to clarify its estimate of the project's on-peak generation. Indeed, without explanation, FERC determined that PacifiCorp *underestimated* the proportion of

Klamath's generation during peak hours. FERC states that "given the current estimate of on-peak and off-peak generation, the resulting composite energy rate was assumed to be about $41.50/MWh." This average value for Klamath's power implies on-peak generation 77.5% of the time.

Value of Replacement Power

Under the FPA, when FERC reviews licensing proposals, in addition to assessing power and other developmental uses of the waterway (such as irrigation and water supply), it must give equal consideration to the purposes of energy conservation, the protection and enhancement of fish and wildlife habitat, the protection of recreational opportunities, and the preservation of other aspects of environmental quality. How FERC conducts this "balancing" among purposes is a bit of a mystery, because FERC does not document its work. It is not that FERC does not monetize the non-developmental/non-market values, which it does not; it usually does not even quantify these values. No matter what the method is that FERC uses to compare and balance the various purposes of the waterway and arrive at a decision in the public interest, it is clear that the price which FERC adopts to value the power generated by the hydro facility plays a crucial role in the decision-making process. This price is used to calculate the power benefits of the project, and it is these power benefits which FERC weighs against all other potential uses of the waterway.

Spot-market prices and the overvaluation of power in the DEIS: FERC staff adopted power values for Klamath of $43.62/MWh (on-peak) and $34.20/MWh (off-peak), citing PacifiCorp as the source for these prices and stating that these values are based on Mid-Columbia and California-Oregon border spot-market prices. The adoption of these prices constitutes a blatant overvaluation of Klamath's power and a shocking acknowledgement by FERC that it used spot-market prices. The power value is supposed to reflect the least-cost of replacing the power which would otherwise have been generated by the Klamath Project. That is, if for some reason a relicensed Klamath Project is unable to generate the power we want to know the cheapest way of replacing that lost power on a sustained basis. No one would rely on the spot market to purchase power for a sustained period of time.

The Department of the Interior (DOI) points out that one purchases power in the spot market only in emergencies (when demand for power was unanticipated) or when the spot market is the cheapest source (when demand was expected, but the amount of power needed to meet the demand is so small that the cost of planning and arranging for a source in advance exceeded the spot-market price). FERC is well-aware of this. That it accepted the power values submitted by PacifiCorp without comment, qualification, or adjustment is beyond comprehension.

Spot markets are very volatile. In tight markets when supply is limited, spot prices are likely to exceed the cost of purchasing firm power for a sustained period. The reverse is the case when there is excess supply, with spot prices covering energy or operating costs and making no or minimal contributions toward capital costs.

As noted above, the power value is supposed to reflect the lowest cost of replacing the power which would otherwise have been generated by the project. In its comments on the DEIS, DOI observed that FERC had several conventional options at its disposal for valuing Klamath's power:

PacifiCorp's average cost of purchase power : Power companies file Form 1 annually with FERC. Among the data which a company provides is a detailed list of the power which it purchased from other sources. PacifiCorp reported that in 2004, it paid an average price of $23.60/MWh for purchasing power from other providers. This is $17.90 less than the $41.50 that FERC selected as the power value in the DEIS.

Wholesale price for interruptible power: According to DOE/EIA, firm power is "power or power-producing capacity, intended to be available at all times during the period covered by a guaranteed commitment to deliver, even under adverse conditions." The market price for firm power includes both an energy charge (the cost of generating power at the margin...the operating cost) and a capacity charge (the capital cost of the facility used to generate the power). If, as PacifiCorp and several stakeholders contend, Klamath is not a dependable source of power, it should be valued as interruptible power, not firm power, and the price for this power should not include a capacity charge.

PacifiCorp's marginal cost of generation: The marginal cost of generation includes all costs except capital costs, i.e., operating costs, supervision, fuel costs, maintenance, etc... everything except the return of and on capital. DOI reviewed PacifiCorp's 2004 Form 1 to identify a recently constructed gas combined cycle unit, similar to the type of unit that would be used to generate replacement power at the margin, i.e., to replace any generating facility that was being retired. The 237MW Hermiston plant, put into service in 1996, serves as a conservative proxy for a replacement plant. New gas combined cycle plants are more efficient than those put in service 10 years ago, so Hermiston's marginal cost of generation stands as an upper-bound for that of a newly constructed plant. PacifiCorp reports an average production expense of $29.10/MWh for Hermiston.

These three different approaches for estimating the cost of replacement power for the Klamath Project produce remarkably similar results, ranging from $23.60-$29.10/MWh. FERC has used a price ($41.05/MWh) to value Klamath's power which is anywhere from 30-43% too high.

Efficiency Upgrades

Electrical systems incur losses in the process of transmission and distribution (T&D) of power. These losses are of two types: line losses (basically friction) and transformer losses (losses in the process of stepping up and stepping down voltage). Since the mid-1940s there have been significant efficiency gains, with T&D losses being halved from 15% of generation to 7-8%. (Losses during peak periods are considerably higher.) Substantial efficiency gains remain to be realized, however, especially through the upgrading of transformers from steel core to amorphous metal core (AMC).

Line losses amount to about 2-3% of generation, while transformer losses account for 5-6% of generation. Transformer losses are roughly equally divided between losses unrelated to load, known as No Load losses, and Load Losses (losses which occur only when power is being transmitted). AMC transformers can reduce No Load losses by approximately 70%, i.e., from 2-3% of generation to .6-.9% of generation. The economics of this savings is favorable, with capital being repaid in less than 4 years.

PacifiCorp reported T&D losses of 3,447,489MWh against sales of 76,601,839MWh, or roughly an average loss of 4.3 percent. T&D losses of this magnitude suggest that PacifiCorp must still have a significant complement of steel core transformers on its system. DOI conducted an analysis of PacifiCorp's T&D system, and estimated that by substituting amorphous metal core transformers, PacifiCorp's *marginal* transmission and distribution costs would increase from the current $23.11/MWh to $24.31/MWh, an increase of only $1.20/MWh.

Rather than generating additional power to replace the lost Klamath power, PacifiCorp would be conserving power by reducing line losses.

Consider the case of normal maintenance, where a steel core transformer has reached the end of its expected productive life and is scheduled for replacement. By replacing it with an AMC transformer rather than a steel core transformer, PacifiCorp realizes efficiency gains and acquires additional power at the cost of $1.20/MWh. The amount of power conserved depends on where in the system the new AMC transformer is located. Transmission losses would be reduced from PacifiCorp's stated 4.5% to about 2.8%; primary distribution loses from 6.9% to 4.8%; and secondary distribution losses from almost 10% percent to 7.9 %.

Note that the $1.20/MWh cost for replacement power derives from a scenario in which the transformer in question is already scheduled for replacement. If early retirement of transformers is involved, then the analysis becomes more complicated, and the cost increase depends upon how early in its productive life the transformer is retired.

Clearly, at the margin, conservation through the replacement of steel core transformers with AMC transformers is among the cheapest way to replace lost power from the Klamath Project. $1.20/MWh is the lowest figure among the options we have examined, and $24.31/MWh falls at the low end of the range for the other options, e.g., purchasing power ($23.60) or generating it ($29.10).

The point here is not that conservation through upgrading of transformers is the cheapest option. The point is that there is enough information to suggest that it is an option deserving of serious consideration. DOI provided its analysis to FERC on March 27, 2006, fully 6 months prior to the issuance of the DEIS. FERC has never acknowledged or responded to DOI's analysis, and did not address this option in the DEIS.

PART-IV

ENVIRONMENTAL IMPACTS OF HYDROPOWER DAMS

12

Impacts of Hydro Dams on Environment[1]

Roger Young, Graeme Smart and Jon Harding

Editor's Introduction: This paper from New Zealand narrates various environmental impacts of dams that we mostly ignore.

There are two ways of storing river water. 'On-line' storage involved holding back water within the river channel by a dam. 'Off-stream' storage involves diverting water away from the river channel for storage. The environmental impact of off-stream projects is less severe because the functions of the river are left unaffected—only the quantity of water is reduced during removal periods. We in India have only made on-line dams. There is a need to shift to off-stream storage. Dams like Tehri and Sardar Sarovar can be decommissioned by building such off-stream storages for providing water during lean periods. Off-stream projects are, of course, more expensive. But we need to examine the costs and benefits dispassionately.

The author says that large fluctuations such as 20 meters in Lake Hawea are 'no longer considered acceptable'. Large fluctuations cause erosion of shorelines, kills vegetation along shorelines and reduces light available to aquatic life. In India we do not appear to acknowledge such impacts.

Sediment accumulated in reservoirs reduces the ability of the river to handle flood flows and increases risk of flooding of upstream cities. The bed level upstream of the Waihopai dam has increased by 25 meters since 1926.

Sediments are flushed from ROR hydropower dams periodically. Indian experts claim that the total flow of sediment in the cycle, of, say, 15 days, is not affected hence there will be no impact on the ecological function. This paper rebuts this view. It points out that erosion takes place when the sediment level is low and deposition takes place when it is high. Thus the downstream areas are, at the least, subjected to repeated bouts of erosion. There is no guarantee that deposition will take place at the place of erosion. Deprivation of sediments threatens the stability of downstream bridges and other structures.

Embankments made for flood control lead to deposition of sediments in the river bed and raise the level of the same. The devastating impact of this was seen in the Kosi floods in Bihar in 2008. They also lead to lesser recharge of ground water and lowering of ground water levels.

A dam 'resets' the water conditions below the impoundment. The water begins to regenerate as it flows again. Thus the quality water due to Tehri dam is 'reset' at discharge and regenerates as it flows freely once again. However, such regeneration will not take place when we build a cascade of dams and no free flow of water is obtained downstream.

The pressure inside turbines may force large quantities of gasses like methane into water in form of bubbles. These bubbles can get into various parts of the body. Major fish kills have resulted from such super-saturation of gas. Consumption of downstream water can have similar effect on human beings.

Changes in flow of water between peak- and non-peak hours have a decisive impact on aquatic life. Fish may find an area suitable during peak hours but not during non-peak hours. They may not be able to adjust to the changes. Further the invertebrates which are food for the fish may not be able to adjust to these changes.

Embankments on low-gradient floodplains have large impact on aquatic life because they deprive fish from access to large flood areas.

The paper speaks of 'self-essence' of a river—cultural values and aesthetics. These are degraded by making of dams. This self-essence is same as the veneration of rivers as Mother in our tradition.

The increase in recreational activities such a boating in hydropower reservoirs is offset by loss of other recreational activities such as rafting and kayaking. These aesthetic values are determined by culture of the people. Reductions in flow will influence the 'sound, smell and overall mood' of the river.

Introduction

Many New Zealand rivers are modified by large water projects. Water is used for irrigation, water supply and industrial purposes throughout New Zealand, but the amounts of water abstracted are often relatively small (less than 1 % of New Zealand's total water resources; Statistics New Zealand 2000). Large-scale abstraction and storage of water for irrigation or water supply occurs primarily in drier parts of the country, or close to larger cities and industrial areas. In contrast, hydro-electric power schemes use 16-33% of New Zealand's total water resources (Statistics New Zealand 2000), but generally allow water to be returned to the same river system. Water storage is often a key feature of large water projects. Storage can be "online", where water is held back within the river channel by a dam, or "off-stream", where water is diverted to a reservoir away from the river channel. The effects of large water projects depend to a great extent on the type and volume of storage involved. Online storage requires the construction of large dams and interferes with many natural processes within the river channel. In contrast, off-stream schemes might be expected to have more benign effects on the environment, but such schemes are uncommon due to the difficulty in locating suitable water storage areas away from the river channel and the costs of transferring water out of the river channel.

Other large projects associated with river systems include flood protection and river control works. Here floodwaters and channel movement are confined to a restricted area to protect the surrounding land from flooding and erosion. Such projects have been instrumental in providing sufficient security for the development of/and within river floodplains and have resulted in significant economic benefits to New Zealand. However, confinement of rivers can have negative effects on the ecological functioning of river systems, reducing the amount of available habitat and also altering natural geomorphic processes.

This paper summarises the effects of large water projects on flow regimes, lake levels, sediment movement and water quality. Changes to the life-supporting capacity of habitats affected by large water projects are also examined, along with their effects on cultural, aesthetic and recreational values that are cherished by many New Zealanders. We also provide a short summary of how the impacts of large water projects are managed within the context of the Resource Management Act (RMA).

Downstream Flow Regimes

Large water projects have caused dramatic changes to the flow regimes of many New Zealand rivers. The diversion of water to Doubtful Sound as part of the Manapouri Power Scheme, for example, has substantially reduced flows in the Waiau River downstream of Lake Manapouri, while diversion of water from the headwaters of the Vhanganui and Rangitikei rivers, as part of the Tongariro Power Development, has increased flows in the Waikato River. Water storage reservoirs tend to attenuate flood peaks and change the annual distribution of flows by storing water at one time and releasing it at another. Flows in the Waitaki River, for example, are highly moderated by the presence of control structures on Lakes Tekapo and Pukaki. These lakes are capable of absorbing large floods and generally augment flows in winter when they would naturally be at their lowest (Waugh and Payne 2003). Fluctuations in the spot price of electricity mean that generation generally peaks for a few hours during weekday mornings and again in the late afternoon when demand is highest. If storage is available, water is held back during weekends and at night. These fluctuations in demand and generation result in substantial fluctuations in flow over short periods.

Lake Levels Upstream

Fluctuating water levels are a feature of all lakes and are controlled by a combination of factors, including catchment size, topography, climate, lake size and the characteristics of the lake outlet. In natural lakes, levels generally fluctuate by less than 5 m, while the levels of storage lakes controlled by large water projects tend to fluctuate more widely (Mark 1987; Fig. 12.1). For example, Lake Hawea historically fluctuated by around 3 m prior to control, and then by over 20 m after the lake was controlled (Mark 1987). Such large fluctuations in lake level are no longer considered acceptable, and the normal operating range is now restricted to 8 m (Freestone and Payne 2000).

Fluctuations in lake levels can have a variety of effects. High lake levels can cause erosion of shorelines, the inundation and mortality of terrestrial vegetation around the shore, and a decrease in available light to littoral and benthic aquatic plant communities (James *et at.* 2002). Low lake levels can also increase rates of lake shore erosion and expose littoral

Fig. 12.1 The Cobb Reservoir after a prolonged period of dry weather. Photo: Roger Young.

communities to freezing, dessication and wave action (James *et at.* 2002).

Storage dams also have the potential to increase the risk of flooding for communities upstream. For example, the presence of the Roxburgh Dam, and sediment deposited in the upper reaches of Lake Roxburgh, appear to have increased the risk of flooding in the town of Alexandra just upstream (Mackay *et at.* 2000). To address this problem, the level of the lake is lowered prior to floods and increased water velocities are used to flush sediment further down the lake. Some sediment is also carried downstream past the dam. This flushing program and the practice of drawing down the lake in advance of a flood has combined to reduce flood levels at Alexandra by approximately 1.7 m (Mackay *et at.* 2000).

Morphologic Effects

The largest New Zealand rivers move millions of tons of sediment to the sea each year. Changes brought about by large water projects can significantly modify sediment transport, and hence river geomorphology (Fig. 12.2). The size of river sediment ranges from suspended clay particles that remain in the water column, to larger substrate materials that move along the riverbed. The finer the particles, the further they travel after being entrained into the river flow. In a flood, entrained clay particles may travel through the river system and out to sea, whereas gravel may only be transported from one river bar to the next. Thus, water projects can have both local and remote morphological effects.

The total sediment load carried by a river is often evaluated in terms of the suspended load and the bed load. At a specified location and flow there is an upper limit to the bed load that a river can transport. Scour and erosion can take place when the actual bed load is below the local transport capacity. Deposition occurs if the transport capacity is exceeded. Suspended load can be deposited in backwaters, on floodplains or within voids between bed particles. The types and rates of sediment erosion and re-deposition determine the morphology of a river system. Rivers can be single thread, meandering, braided or a sequence of steps and pools. Morphology is relatively stable when the long-term supply of sediment to a river is matched by a river's ability to transport sediment, but water projects can alter the balance between sediment supply and transport capacity and bring about serious morphological changes.

Dams can result in improved water clarity downstream, as their reservoirs trap most of the sediment that previously passed through. Sediment trapping can cause adverse effects both upstream and downstream. Sediment deposition within the reservoir can reduce the intended storage volume and may raise the riverbed upstream. For example, the Waihopai Darn in Marlborough has trapped 10 million cubic metres of sediment since its construction in 1926 and increased bed levels upstream by up to 25 m. Downstream of a dam the combination of clear

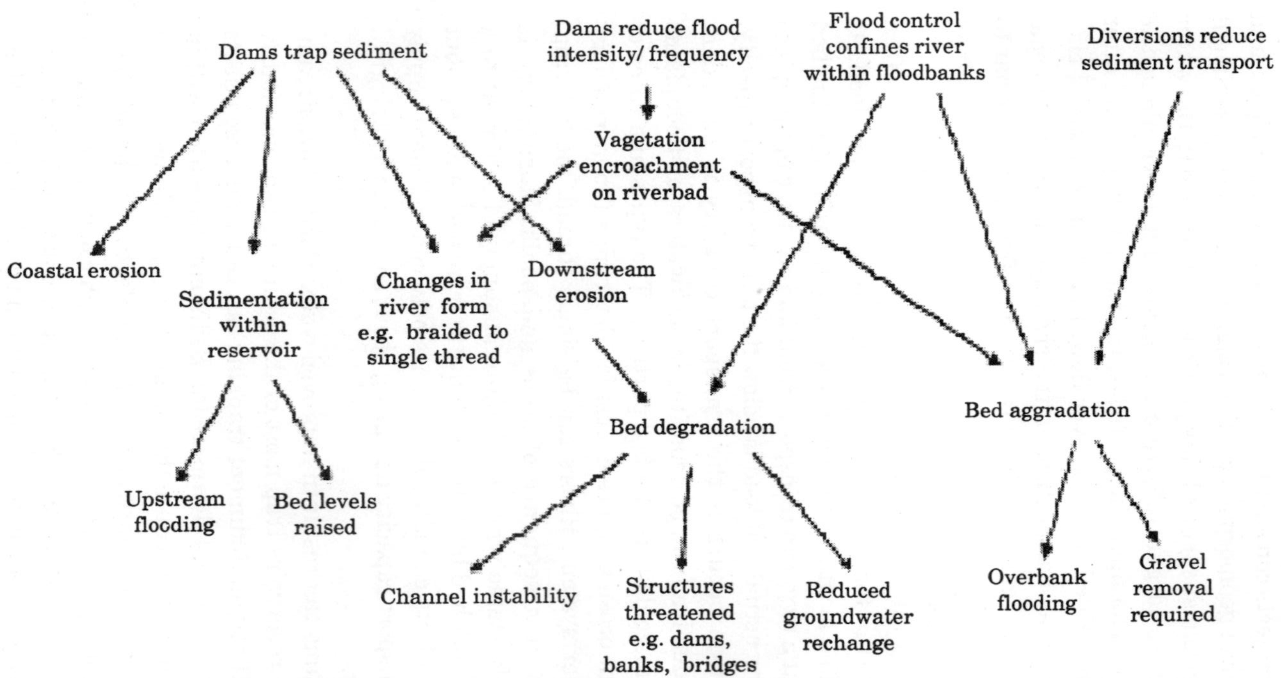

Fig. 12.2 Conceptual diagram of the potential effects of large water projects on river morphology

Fig. 12.3. Piles exposed beneath the piers of the Claude- lands Rail Bridge on the Waikato River, Hamilton.

water and modified flow patterns can increase erosion and lower the rivet bed. Lowering of the river bed may endanger the structural stability of the dam, undermine banks and bridges and lower groundwater levels. Changes to the channel will eventually cease when the downstream river slope becomes flatter or when the bed becomes armoured with stones large enough to resist erosion. Fig. 12.3 shows bridge piles that have been exposed as a result of falling bed levels in the Waikato River. Parts of this river are degrading because sediment supplies from upstream are trapped behind dams, and sediment starvation allows the river to scour and transport material from the bed and banks downstream of the dams.

Water diversion intakes can also affect the sediment regime of a river. If too much water is taken, the flows downstream of an intake may not be sufficient to move incoming sediment and sediment deposition will occur, raising the bed level. Channel instability and over-bank flooding may also occur. Sediment may also be diverted along with the water, causing problems. This occurred at the Poutu Intake of the Tongariro Power Development in the late 1980s, when 42,000 tons of sediment per year were diverted from the Tongariro River and had to be excavated from canals.

Islands within river systems are particularly susceptible to changes associated with flow regulation. For example, changes in sediment dynamics may cause island growth, migration or erosion. Some low-lying islands may be drowned when river depths increase. Conversely, if flows and depths are reduced, previously inaccessible islands may be easily accessed by the public and by introduced predators. This is a particular concern for local iwi, since many riverine islands are waahi tapu or offer sanctuary to rare species (MfE 1998).

Effects on Coastal Erosion

Where a dam intercepts the natural export of sediment to the ocean, severe effects can be seen in the coastal zone, as beaches may be depleted and coastal erosion accelerated. Concerns have been raised about the effects of sediment trapping by the large dams on the Clurha and Waitaki rivers, since both rivers historically contributed significant amounts of sand and gravel to the Otago and South Canterbury coasts. The potential yield of sand and gravel from Clutha River tributaries to the coast has been reduced by 95% since the Roxburgh and Clyde dams have been in place (Hicks *et al.* 2000). However, the effects of this reduction on rates of coastal erosion are not clear. Rates of coastal erosion north of the Waitaki River mouth appeared to increase immediately after the construction of the Waitaki Dam, but have now returned to pre-dam levels (Hicks *et at.* 2002).

Geomorphic Effects of Flood Control Schemes

By protecting land from inundation, flood control schemes prevent flood flows from covering parts of their original floodplain. Floodwaters and their associated sediment load are concentrated into a channel or floodway. Not only is sediment that would have been deposited on the floodplains conveyed further downstream, but also the concentrated flows have a greater ability to erode and transport sediment. Local degradation and downstream sediment accumulation can occur. Narrowing of braided rivers sometimes has the opposite effect and causes rapid aggradation in the riverbed between the floodbanks. Because of this process, reaches of the North Ashburton River are now perched above their original floodplain. Substantial aggradation of riverbeds between floodbanks decreases the level of flood protection offered by the banks, while substantial degradation has the potential to undermine the foundations of the floodbanks, thus requiring expensive bank protection works. Riverbed levels also influence the rate of recharge to surrounding aquifers. For example, a 0.5 m drop in the lower Motueka riverbed is predicted to reduce summer recharge of the Motueka Plains aquifer by 24% (Basher 2003). Changes in the roughness of channel banks, such as occur when replacing scrub-covered banks with grass or rock, can also bring about large changes in the sediment-transporting characteristics of a river channel (Yu and Smart 2003).

Effects of Storage on Water Quality

Storage of water in reservoirs often results in changes to the physical and chemical characteristics and temperature of the water released downstream (Table 12.3). The changes depend on a variety of factors, including the residence time (i.e., the average length of time taken for water to pass through the reservoir), the position of the reservoir within the catchment (i.e., in the headwaters or lower reaches), whether the water within the reservoir becomes stratified and anoxic, and the level of the reservoir outlet (surface or deep).

Generally, the longer the residence time, the greater the potential effects of water storage. As mentioned above, most of the suspended sediment washed into a deep reservoir will be

Table 12.3

Changes in downstream water quality associated with water storage

Parameter	*Change*	*Comments*
Suspended Solids	Reduced	Most marked effects in large deep lakes. Sediment may be re-suspended by wind in shallow lakes.
Water clarity	Increased	Dense blooms of phytoplankton in lakes may reduce clarity at times. Release of iron and manganese in some reservoirs may reduce water clarity downstream.
Organic matter	Reduced	However, high quality organic material is often released from reservoirs.
Water temperature	Reduced daily temperature fluctuations. Increased summer daily mean temperatures in headwater reservoirs. Seasonal buffering of temperatures in large lowland reservoirs. Delayed seasonal minima and maxima.	
Dissolved oxygen	Reduced	Depends on the degree of stratification.
Phosphorus, Iron and Manganese	Increased	Particularly if bottom waters of reservoir are anoxic.
Nutrients	Increased	Especially after inundation of fertile soils by a new reservoir.

trapped within the reservoir. Therefore water released downstream will generally be clearer, with lower concentration of suspended sediment. The classic examples of this in New Zealand are the Roxburgh and Clyde Dams, which intercept much of the sediment from the Kawarau and Showver rivers upstream. As well as geomorphic effects, the changes in water clarity and suspended sediment concentration can have marked effects on the ecology of river systems downstream.

Organic matter (e.g., leaves, wood and other detritus) can also be trapped within reservoirs, altering the composition and delivery of material to downstream reaches. Organic matter from upstream is often a key food source for many riverine organisms (Fisher and Likens 1973). The interception of organic matter may interfere with the natural longitudinal changes in river ecosystems and the productivity of downstream reaches

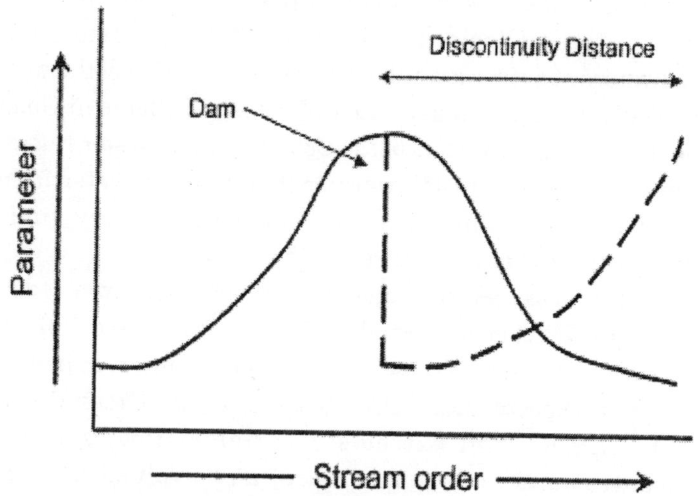

Fig. 12.4. Conceptual diagram of the effect of dams in resetting biotic and abiotic parameters compared to a natural river system (redrawn from Ward and Stanford 1983)

(Vannote *et al.* 1980). Ward and Stanford (1983) proposed the Serial Discontinuity Concept, which suggested that natural and artificial lakes "reset" many abiotic and biotic conditions below impoundments (Fig. 12.4). The downstream or upstream shift of a particular abiotic or biotic parameter caused by impoundment is defined as the discontinuity distance, while the magnitude of change is referred was the parameter intensity (Fig. 12.4). The construction of a dam might be expected to reduce biodiversity in the downstream river, for example, and maximum biodiversity may not return to pre-impoundment levels for many kilometres downstream.

A reservoir near the headwaters of a river catchment is predicted to have markedly different effects than one in the lower reaches (Ward and Stanford 1983). For example, the water temperature of forested headwater streams is generally cool throughout the year due to shading by overhanging riparian vegetation and the buffering influence of groundwater contributions (Poole and Berman 2001). A dam in the headwaters of a river may allow more solar heating of the water, especially during summer. The Cobb Power Scheme in Golden Bay dams the headwaters of the Cobb River and water released from the dam is up to 4°C warmer than riverine water from a neighbouring catchment during summer, but very similar during winter (Young *et at.* 2000; Fig. 12.5).

In contrast, a large-volume reservoir in the lower reaches of a river will act as a thermal buffer. Water released will often be cooler in the summer, and warmer in the winter, than would have been expected naturally. Historically, the Green River in Utah, U.S.A., had extremely variable flow, sediment concentrations and temperature, and evolved a unique assemblage of native fish species adapted to these extreme conditions. The construction of the Flaming Gorge Dam resulted in major changes to the thermal regime of the river downstream, with much cooler temperarures in the summer and slightly warmer temperarures in the winter. This effect is exacerbated, as the water released downstream is from deep in the reservoir. Many of the native fish species that relied on warm summer temperatures as cues for spawning have been displaced downstream. Introduced cold-water species, such as rainbow trout, now flourish downstream of the dam (Filbert and Hawkins 1995). Such dramatic changes in thermal regimes have not been documented in New Zealand hydro dams. However, significant buffering of temperatures is expected to occur below large natural and hydro lakes.

Dissolved oxygen concentrations in water bodies are controlled by the balance between the supply of oxygen from the atmosphere and photosynthetic inpurs, and the consumption of oxygen by biological and chemical oxidation processes. Oxygen exchange with the atmosphere is reduced in deep reservoirs relative to rivers, and may result in the deeper parts of reservoirs becoming anoxic especially if the water column becomes thermally stratified. In addition organic

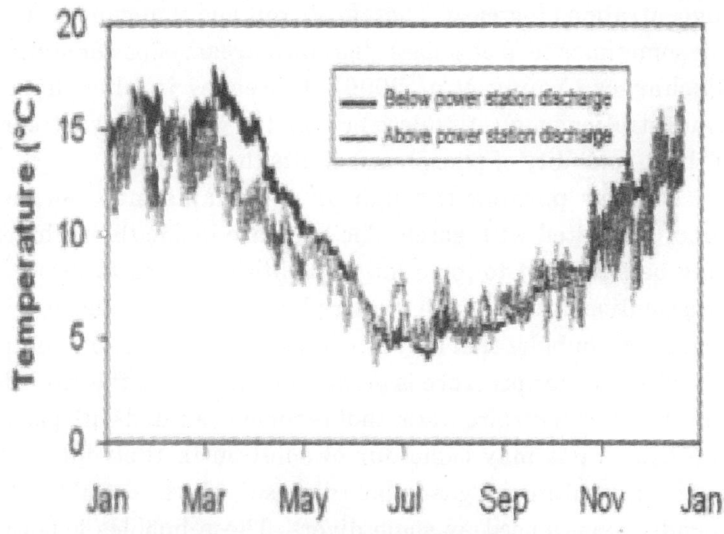

Fig. 12.5. Changes in the annual water temperature regime as a result of storage in the headwaters of a river catchment (Takaka River, Golden Bay).

matter will tend to sink to the bottom of reservoirs, where it eventually decomposes, increasing the demand for dissolved oxygen. Water released from the deeper parts of reservoirs may be depleted in dissolved oxygen, with direct effects on biota downstream. However, exchange with the atmosphere will eventually bring oxygen back to equilibrium concentrations. This may happen very quickly if water is passed through turbines, resulting in turbulent mixing of water and air.

The concentration of dissolved oxygen near the bottom of a reservoir also influences the concentrations of phosphorus, iron and manganese in the water (Wetzel 1983). As the concentration of dissolved oxygen declines near the sediment surface, the release of phosphorus, iron and manganese increases markedly. The change in redox conditions, caused by low oxygen concentrations, results in the reduction of insoluble iron and manganese hydroxides to more soluble forms, and the mobilization of adsorbed phosphorus. Dissolved iron and manganese will precipitate out of solution once oxygen concentrations increase. Therefore, iron and manganese flocs are sometimes seen on substrates downstream of power station discharges (Young *et al* 2000). Increases in phosphorus concentrations may stimulate periphyton growth downstream if the availability of phosphorus limits their growth.

Water passing through or over dams may become super-saturated with gases. The pressure inside the turbines may be sufficient to force large quantities of gas to dissolve into solution. Alternatively, water plunging over a spillway will contain air bubbles that may be carried to a considerable depth, where the water pressure is great enough to force the gas into solution. If fish respire water that is supersaturated with gases, the excess gas may come our of solution in their blood as bubbles, resulting in "gas-bubble disease", which resembles the "bends" experienced by some divers. These bubbles lodge in various parts of the body and can cause injury or death. Major fish kills have resulted from this super-saturation below several dams overseas (Backman *et at.* 2002). To our knowledge. mortality due to gas supersaturation has not been documented in New Zealand.

Effects of Diversion on Water Quality

Diversion of water from one catchment to another also results in the diversion of sediment, nutrients and organic matter. Therefore, any benefits or problems resulting from this material may also be transferred berween catchments. Nutrient availability often determines phytoplankton production around parts of New Zealand's coast. Therefore, the productivity of some coastal areas around river mouths may depend on the supply of nutrients and organic matter from adjacent rivers, and could be affected if this supply is diverted elsewhere. This concern has been raised in relation to the Manapouri Power Scheme's diversion of Waiau River water to Doubtful Sound, and the potential effect on the ecology of the Southland coast.

Diversion of water also reduces the ability of a river to dilute wastes. This is particularly a problem with irrigation schemes, where the diverted water allows more intensive agriculture in the surrounding catchment. Runof from this irrigated land then makes its way back into the river and is usually highly enriched with nutrients and other contaminants. The effects of diversion are essentially doubled, since the reduced flow of the river is less able to dilute these elevated levels of contaminants in the runoff.

Reductions in flow may also alter dissolved oxygen and temperature dynamics in rivers. Lower, more stable flows and slower velocities tend to allow higher biomasses of periphyton to accumulate. Assuming that more biomass means more photosynthesis, the amount of oxygen released during daylight and taken up at night by periphyton mats will also be increased. Under low flow conditions, this increased flux of oxygen will be dissolved in a smaller volume of water, potentially resulting in larger daily fluctuations in dissolved oxygen concentration if the amount of exchange of oxygen through the river surface is not altered. The thermal buffering effect of a large water volume will also be reduced at lower flows, potentially leading to larger daily fluctuations in water temperature.

Ecological Effects of Lake Formation

The impoundment of rivers and creation of artificial lakes causes dynamic shifts in pelagic and benthic communities in these new water bodies. Over time the community composition changes, with a reduction in lotic (running water) species and an increase in lentic (still water) ones. The formation of Lake Dunstan in Central Otago in 1992 resulted in the flooding of surrounding river terraces, which caused significant changes in energy transfer within the new lake, as terrestrial vegetation and organic material associated with the land decomposed. During the early years of the lake, the benthic fauna was dominated by chironomids and oligochaetes. However, as this source of enrichment dissipated, and aquatic macro-phyces invaded, the community changed to one dominated by snails and caddis flies (Strickland *et al.* 2000). New reservoirs will usually be invaded within a few years by aquatic macrophytes. Lake Dunstan had not been completely filled when *Elodea canademis, Myriophyllum spp.* and *Lagarosiphon major* were observed in the lake. By 1998, six years after the dam have been completed, a stable community of 16 aquatic plant species had formed throughout the lake. *Lagarosiphon* dominated in the 2-4 m zone (Strickland *et al.* 2000).

Not surprisingly, the formation of a new lake will create habitat for aquatic birds. Eight years after Lake Dunstan was filled, 25 species occurred around the lake, compared with 14 recorded prior to lake formation. New colonisers included the New Zealand scaup, Canada geese, Black swans, Australian coots, and Black and Little shags. Lake-level fluctuations resulting from changes in hydro-electric power generation have positive and negative effects on bird communities. The most important impacts are on nesting activities of species such as grebes, coots, pukeko and black swans, which construct floating nests or nest close to the shoreline. Feeding can also be affected, as the raising and lowering of lake levels may reduce the availability of aquatic plants and benthic invertebrates as food, or alternatively make terrestrial food that was previously flooded available (Sanders 1996). However, these effects vary markedly with the lake morphology and riparian vegetation of the lake deltas.

Effects on People

People value waterways for a variety of reasons. A river may have important cultural values (mauri, mahinga kai, waahi tapu), be important for recreation (fishing, whitebaiting, kayaking, rafting, swimming, jet boating), and/or make a major contribution to the aesthetics of the surrounding landscape (MfE 1998). Large water projects have the potential to degrade these values if safeguards are not put in place. For example, flow regulation changes the seasonal pattern of flows, the strength of the connection between the mountains and the sea, and the ability of a river to carry sediment and dilute contaminants. All these changes will affect the mauri or life-essence of a river. Diversions resulting in the mixing of water from one catchment with another will also desecrate the mauri of rivers (MfE 1998). River-mouth blockages associated with reduced flows, along with dams themselves, can adversely affect the passage of mahinga kai species between spawning and rearing areas. Many waahi tapu are located near rivers and lakes, and can be inundated by storage dams, or overgrown with weeds due to reductions in flood frequency caused by river control upstream. Kai Tahu considers that an estimated 90% of waahi tapu sites adjacent to Lake Hawea, Lake Dunstan and Lake Roxburgh have been lost because they were sited next to the river and are now drowned under the hydro-electricity lakes (Contact Energy 2001).

Many of New Zealand's large water projects have improved public access and created reservoirs, which are recreational assets for anglers, swimmers and boat owners. The network of lakes in the Waitaki Valley, for example, attracts numerous holidaymakers every summer. However, these improvements in recreational facilities have often been offset by the loss of recreational activities using the flowing waters inundated by reservoirs. For example, rapids on the Kawarau River that were popular for rafting and kayaking have been inundated by Lake Dunstan (Egarr and Egarr 1981). Changes in flow regime may also influence the quality of the recreational experience on some waterways. Reductions in water depth may expose debris. such as logs and rocks, posing serious risks for swimmers and paddlers. The safety of some swimming holes/ jumping sites that require deep water may also be affected by reduced flows. Changes in water quality as a result of large water projects may also affect people's enjoyment of an area. An interesting example is the Tekapo River, where water clarity has increased now that turbid glacial water is diverted down the hydro-electric canals. While this increase in clarity and reduction in flow fluctuation has improved the trout fishery in the Tekapo River, local iwi are concerned that the stabilised flows have benefited salmon ids and thus increased predation and competitive pressures on native fish species (MfE1998).

Aesthetic values are determined by the landscape's physical and natural properties, coupled with the cultural values of the person experiencing the landscape (MfE 1998). Large water projects may affect the physical properties of a river by directly altering flow variability, water colour and clarity, hydraulics (width, depth and velocity), and the amount of riverbed exposed during low flows. Indirect physical changes associated with large water projects include modifications to riverbed vegetation and channel type (braided versus single thread) caused by changes in the sediment supply or flooding frequency. The extent of physical effects on aesthetic values will depend on the channel shape, with the most apparent effects on single-thread rivers with gently shelving margins. Very large changes in flow are required to change the wetted area of a river with near-vertical banks, while braided rivers are naturally dominated by exposed gravel banks, so additional exposed areas are less conspicuous (MfE 1998). Other perceptual qualities of a river may also be affected by large water projects. Artificial structures, like dams and canals, will reduce the natural character associated with a river, while reductions in flow may influence the sound, smell and overall mood of a river. The wild and scenic properties of a large, raw, untamed river will be degraded by flow regulation and reduction (Godman 1989).

References

BACKMAN, T.W.H.; EVANS, A.F.; ROBERTSON, M.S.; HAWBECKER, M.A 2002: Gas bubble trauma incidence in juvenile salmon ids in the Lower Columbia and Snal(e rivers. Noreh American Journal of Fisheries Management 22: 965-972.

BASHER, L.R. 2003: The MotUeka and Riwaka catchments - a technical report summarising the presenr state of knowledge of the catchment, management issues and research needs for inregrated catchmenr management. Landcare Research NZ, Lincoln.

CONTACT ENERGY LTD 2001: Application for Resource Consents and Assessment of Environmental Effects -Clutha Catchment Hydro-electric Generation Facilities.

EGARR, G.D.; EGARR, J.H. 1981: New Zealand Recreational River Survey. Part III South Island rivers. Water and Soil Miscellaneous Publication No. 15. National Water and Soil Conservation Organisation.

FILBERT, R.B.; HAWKINS, CP. 1995: Variation in condition of rainbow trout in relation or food, temperature, and individual length in the Green River, Utah. Transactions of the American Fisheries Society 124: 824-835.

FISHER, S.G.; LIKENS, G.E. 1973: Energy flow in Bear Brook, New Hampshire: an integrative approach to stream ecosystem metabolism. Ecological Monographs 43: 421-439.

GODMAN, L. 1989: The Last Rivers' Song - Photographs of the Clutha and Kawarau. John McIndoe Limited, Dunedin.

HICKS, D.M.; WALSH, J.M.; DUNCAN, M.J. 2000: Clutha River sediment budget. NIWA client report CHCOO/45.

HICKS, D.M.; WILD, M.; TODD, D. 2002: Project Aqua: Coastal and river mouth effects. NIWA client report CHCOl/112.

JAMES, M.R.; MARK, AF.; SINGLE, M. 2002: Lake Manager's Handbook: Lake level management. Ministry for the Environment, Wellington.

MARK, A.F. 1987: Lake level conrrol. In: Henriques, P.R. (ed.) Aquatic Biology and Hydroelectric Power development in New Zealand. Oxford University Press, Auckland, 113-123.

MACKAY, CA.; WEBBY, M.G.; WALSH, J.M. 2000: Lake Roxburgh hydraulic modelling. Report for Contact Energy Ltd., Opus International Consultants, Wellington.

MfE 1998: Flow guidelines for instream values. Volumes A and B. Ministry for the Environment, Wellington.

POOLE, G.C; BERMAN, CH. 2001: An ecological perspective on in-stream temperatUre: natural heat dynamics and mechanisms of human-caused thermal degradation. Environmental Management 27: 787-802.

SANDERS, M.A 1996: Effects of fluctuating lake levels and habitat enhancement on black stilts *(Himantopus novaezefandiae* Gould, 1841). Ph.D thesis, University of Canterbury, Christchurch.

STATISTICS NEW ZEALAND 2000: New Zealand Official Yearbook 2000. Statistics New Zealand, Wellington.

STRICKLAND, R.R.; HARDING, *J.S.;* SHEARER, K.A. 2000: Biology of Lake Dunstan. Prepared for Contact Energy Ltd. Cathron Report No. 563.

VANNOTE, R.L.; MINSHALL, G.W.; CUMMINS, K. W; SWEIL, J.R.; CUSHING, CE. 1980: The river continum concept. Canadian Journal of Fisheries and Aquatic Sciences 37: 130-137.

WARD, J .Y.; STANFORD, J.A 1983: The serial discontinuity concept oflotic ecosystems. In: Fontaine, TO.; Bartell, S.M. (eds.), Dynamics of Lotic Ecosystems, Ann Arbor Science: 29-41.

WAUGH, J.R.; PAYNE, D.A. 2003: Projecr Aqua Waitaki River Hydrology Study. Opus International Consultants, Wellington.

WETZEL, R.G. 1983: Limnology. Saunders College Publishing, Philadelphia.

YOUNG, R.G.; HARDING, J.S.; STRlCKLAND, R.R; STARK, J.D.; SHEARER, K.A; HAYES, J.W 2000: Cobb power scheme resource consent renewal: Water quality, aquatic habitat and fisheries. Prepared for TransAlta New Zealand Ltd, Cawthron Report No. 554.

YU, G.; SMART, G.M. 2003: Aspect ratio to maximize sediment transport in rigid bank channels. Journal of Hydraulic Engineering, ASCE, 129.

13

Impact of Hydro Turbines on Water Quality[1]

United States Court of Appeals

Editor's Introduction: This judgment by a U.S. Court denied license to change the configuration of turbines because it will lead to change in timing of discharge of water. Our Central Electricity Authority approves ROR dams 'with pondage' and allows them to modify discharge of water to meet peak load demands. CEA assumes this does not constitute change in flow of river. This judgment holds to the contrary.

The question raised was whether flow of water that is different in volume, timing, intensity and DO constitutes a change in 'discharge'. The FERC argued that change in timing

1. Judgment in Alabama Rivers Alliance, American Rivers, Inc. And Lake Watch of Lake Martin V. Federal Energy Regulatory Commission, District of Columbia Circuit Court, Decided April 11, 2003, No. 01-1408.

does not amount to 'change in discharge.' The total discharge during the day would be the same. The question was whether higher discharge at particular times such as peak hours and lower discharge at non-peak times amounts to change in discharge. The Court held that change in discharge takes place if the timings are changes even though the "total daily volume" remains unchanged. The Court asks, why *any* discharge should be measured on a daily, as opposed to on a weekly or even yearly, basis? This is opposite to the view of CEA which holds that 'ROR with pondage' projects do not change the flow of river because water is not stored in one season to be used in another season. Whether flow of water is modified inter-season or intra-day makes no difference. Water flow is modified in both cases.

Another aspect brought to light in this judgment is that U.S. law requires a certification from State Government to the effect that water quality will not be affected. Refusal of such certification by the State Government leads to cancellation of the project. There is a need to build-in a similar provision in our environmental laws because our State Governments are toothless in preventing installation of hydropower projects.

Summary

Petitioners Alabama Rivers Alliance, American Rivers, Inc. and Lake Watch of Lake Martin seek review of the decision of the Federal Energy Regulatory Commission (FERC or Commission) to amend an existing hydroelectric license issued to Alabama Power Company (Alabama Power). The amended license authorizes Alabama Power to replace three existing turbine generators at its Martin Dam Project on the Tallapoosa River with new, more efficient units. The petitioners contend that the Commission erred in issuing the license amendment without first requiring Alabama Power to obtain water quality certification from the State of Alabama. Because we conclude that an increase in the volume of water passing through the dam's replacement turbines "may result in any discharge into the navigable waters" within the meaning of section 401(a)(1) of the Clean Water Act (CWA), we grant the petition for review and vacate the Commission's orders.

The Regulatory Framework

The Federal Power Act (FPA) authorizes FERC to issue licenses "for the purpose of constructing, operating, and maintaining dams, water conduits, reservoirs, power houses, transmission lines, or other project works necessary or convenient for the development, transmission, and utilization of power across, along, from, or in any of the streams or other bodies of water over which Congress has jurisdiction" under the Commerce Clause. These hydroelectric licenses contain certain conditions that FERC deems necessary to improve and utilize the nation's waterways in general and water-power development in particular. Upon "mutual agreement" between the Commission and a licensee, FERC may amend such licenses, which are issued "for a period not exceeding fifty years."

Although the FPA represents a congressional intention to establish 'a broad federal role in the development and licensing of hydroelectric power,' the Clean Water Act (CWA) "has diminished [the FPA's] preemptive reach by expressly requiring the Commission to incorporate into its licenses state-imposed water-quality conditions." FERC's hydroelectric licenses are thus subject to, among other conditions, the requirements of section 401 of the CWA.

Section 401(a)(1) of the CWA provides that "any applicant for a Federal license or permit to conduct any activity which may result in any discharge into the navigable waters, shall provide the licensing or permitting agency a certification from the State in which the discharge originates or will originate." The required certification must provide that such discharge will comply with the applicable water quality standards of the CWA, as well as with "any other appropriate requirement of state law." Any limitations included in the state certification become a condition on the federal license. If the "originating" state denies an applicant section 401(a)(1) certification, FERC may not issue that applicant a hydroelectric license.

The License Amendment Proceedings

Located on the Tallapoosa River in central Alabama, Alabama Power's Martin Dam Project has the capacity to generate 154.2 megawatts (MW) of electric power. The dam project generates electricity by taking in water from the reservoir above the dam, funneling the water through four turbine generators and then releasing the water through four penstocks into the river below the dam. The first three 33 MW turbine generators began commercial operation in 1927.

On December 22, 2000, Alabama Power filed an application with the Commission to amend its license to authorize the replacement of the three 33 MW turbine generators. These units had exhibited serious leakage problems and efforts to repair them had been largely unsuccessful. According to Alabama Power's estimates, the replacement turbines would increase the flow of water into the river below the dam project by approximately 900 cubic feet per second (cfs), an increase of 8.6%, and would increase each generator's capacity by 7 to 10 MW, an increase of roughly 20–30%.

Shortly after the Commission provided public notice of Alabama Power's license amendment application, the petitioners moved to intervene, arguing that section 401(a)(1) of the CWA required Alabama Power to obtain state water quality certification before the Commission could amend its existing license. On May 23, 2001, the Commission issued an order approving Alabama Power's proposed license amendment. In doing so, the Commission rejected the petitioners' argument that the license amendment required state water quality certification under section 401(a)(1), concluding that "replacing and upgrading the existing turbine-generator units is not an 'activity which may result in a discharge' within the meaning of section 401(a)(1)."

While the Commission recognized that "replacement of the turbine generators would increase the dam project's hydraulic capacity, with the result that water would be discharged more quickly into the river," it found that "the nature of the discharge would not change," i.e., that "water from the reservoir would continue to be released through the new turbine generators in essentially the same manner as it is now released through the existing turbine generators." Reasoning that section 401(a)(1) requires certification only for "'activities which may result in a discharge,' not activities which may alter an already existing discharge," the Commission held that section 401(a)(1) did not apply to Alabama Power's amended license.

The Commission's Order Amending License likewise rejected the petitioners' argument that 18 C.F.R. § 4.38(f)(7)(iii) required Alabama Power to obtain state water quality certification for its proposed license amendment. Section 4.38(f)(7)(iii) provides that "any application to amend an existing license requires a new request for water quality certification if the amendment would have a material adverse impact on the water quality in the discharge from the project." Citing its environmental assessment of the proposed license amendment, the Commission concluded that the proposed modifications to the three 33–MW turbine generators would not have a "material adverse effect" on the water quality of the dam project's discharge.

Accordingly, the Commission held that section 4.38(f)(7)(iii) did not require Alabama Power to obtain state water quality certification in applying to amend its existing hydroelectric license.

The petitioners moved for rehearing. On July 23, 2001, the Commission denied the motion, again rejecting the argument that section 401(a)(1) required Alabama Power to obtain state water quality certification for its proposed license amendment. The Commission explained that section 401(a)(1) applies only to activities that may "'result in a discharge,'" not to those activities that simply alter an existing discharge. Alabama Power's license amendment did not trigger the certification requirements of section 401(a)(1), the Commission reasoned, because replacement of the existing turbine generators "will at most alter an existing discharge by increasing the amount of water flowing through the units at any given time and, concurrently, reducing the amount of time that the units will operate each day."

Noting that all construction would be completed within the existing structures of the dam project as well as "in the dry," the Commission dismissed the petitioners' argument that replacement of the turbine generators would result in discharge of dust, grease and oil. The Commission rejected the notion that removal of the old turbine generators could result in any such discharge and, in addition, concluded that the possible release of "trace amounts of dust or grease" during the initial installation of the new turbine generators was an insufficient basis to conclude that the replacement would result in a discharge requiring certification under section 401(a)(1). We now review the Commission's orders.

Analysis

On review, the petitioners maintain that the Commission erred in issuing Alabama Power's proposed license amendment without first requiring Alabama Power to obtain a water quality certification from the State of Alabama. Specifically, the petitioners challenge the Commission's conclusion that the installation and operation of the three new turbine generators at the Martin Dam Project is not an "activity which may result in any discharge" for purposes of section 401(a)(1). They insist that the installation and operation of the new turbine generators "may result" in at least two discharges into the Tallapoosa River: (1) an increased flow of water, and particularly of low dissolved oxygen (DO) water passing through the dam's replacement turbines and (2) the release of dust, grease and oil during the installation of the replacement turbines. We agree with the petitioners as to the first of the claimed discharges and, on that basis, grant the petition and vacate the Commission's orders.

Standard of Review

We review a Commission licensing decision under the FPA to determine whether it was "arbitrary and capricious." In reviewing such a decision, we must treat the Commission's findings of fact as "conclusive" if they are "supported by substantial evidence." Because the petitioners did not challenge the Commission's factual findings in their request for rehearing, however, they are precluded from doing so now. We therefore consider the Commission's findings of fact to be conclusive.

As a result, we must simply resolve a question of statutory interpretation—whether the installation and operation of the replacement turbines is an activity that "may result in any discharge" within the meaning of section 401(a)(1) of the CWA. The Commission's interpretation of the CWA is not entitled to the usual judicial deference, however, because the Environmental Protection Agency (EPA)—and not FERC—is charged with administering the statute. We therefore review the Commission's interpretation of section 401(a)(1)—a purely legal question—*de novo*.

The Increased Water Flow

The petitioners' argument is a straightforward one. As previously discussed, section 401(a)(1) requires state water quality certification for "any activity which may result in any discharge into the navigable waters." Emphasizing that water must flow through the turbine generators and be discharged in order for the dam project to produce electricity, the petitioners argue that the licensed activity—*i.e.*, the installation and operation of the replacement turbines—will necessarily result in "any discharge" into the Tallapoosa River. Specifically, the petitioners maintain that the operation of the new turbine generators will result in an increased flow of water "significantly different in volume, timing, and intensity," as well as in the amount of DO, from the flow of water passing through the old turbines. The petitioners thus reason that, under the plain language of section 401(a)(1), Alabama Power's license amendment required state water quality certification.

The Commission contends, however, that the petitioners misinterpret the plain language of section 401(a)(1) and thereby misunderstand what must be shown in order to establish that an activity "may result in any discharge." More specifically, the Commission asserts that the petitioners' argument runs afoul of our reasoning in *North Carolina*, a case in which we considered whether FERC violated section 401(a)(1) by issuing an amended hydroelectric license authorizing a licensee to *decrease* the flow of water passing through a dam's turbine generators without first receiving state water quality certification. Rejecting the argument that section 401(a)(1) requires state water quality certification simply because an activity will "result in an altered discharge," we reasoned as follows:

The existence of certification rights under section 401(a)(1) does not depend on whether a discharge is "altered." Section 401(a)(1) certification rights vests only if an activity "may result in" a discharge. This distinction is of no small moment. The word "alter" means to change something from its previous state implying that the thing changed was already in existence. By contrast, the word "result" implies causation. Obviously, a subsequent event cannot be the cause of something that is already in existence. Given the disparity between petitioners' proposed test and the words of the [CWA], we elect to remain faithful to the language chosen by Congress and require that an activity "result in" a discharge in order to trigger the certification requirements of section 401(a)(1).

Citing "the lack of equivalence between an activity that may *increase* per-second water flows and an activity that '*may result in*' such flows," the Commission contends that our reasoning in *North Carolina* applies with equal force here.

Replacement of the turbine generators cannot, in the Commission's view, possibly "result in" water flows that "pre-existed and will continue, in some form, regardless of whether that replacement is approved." At most, the Commission argues, replacement of the turbine generators will "'alter an existing discharge by increasing the amount of water flowing through the units at any given time and, concurrently, reducing the amount of time that the units will operate each day.'"

Yet the Commission's argument ignores the critical feature of the licensing activity at issue in *North Carolina*: operation of the pipeline project—*i.e.*, the "activity" authorized by the license amendment—resulted in the *withdrawal* of water from the Lake Gaston reservoir. In our view, the distinction between an increased discharge and a decreased discharge is "of no small moment." But this should hardly come as a surprise, for we recognized as much in *North Carolina* itself. Relying on the definition of "discharge of a pollutant" and "discharge of pollutants," we observed that "the word 'discharge' contemplates the *addition*, not the withdrawal, of a substance or substances." Because the operation of the pipeline project would not result in the "addition" of anything to Lake Gaston, we held that "[a] decrease in the volume of water passing through the dam turbines cannot be considered a 'discharge' as the term is defined in the CWA."

The plain language of section 401(a)(1) and our construction of that language in *North Carolina* thus compel us to conclude that Alabama Power's installation and operation of the new turbine generators at its Martin Dam Project is an "activity which may result in any discharge" within the meaning of section 401(a)(1). As discussed above, "the word 'discharge' contemplates the addition of a substance or substances" into the navigable waters. Here, the Commission concluded that the replacement turbines would increase the flow of water into the river by approximately 900 cfs. Thus, at the very least, the replacement turbines will release low DO water into the river at an increased rate of 900 cfs. The installation and operation of the replacement turbines is therefore an activity that "may result in any discharge."

The Commission's reliance on the unchanged daily volume of the discharge is similarly misplaced. While perhaps persuasive at first glance, the Commission's "total daily volume" argument suffers from a fatal flaw: it lacks any basis in the statutory language of section 401(a)(1). To wit, the Commission has offered no argument—persuasive or otherwise—as to why "*any* discharge" should be measured on a daily, as opposed to on a weekly or even yearly, basis. Given the possibility that even a temporary increase in a discharge could have a negative water quality impact, "we elect to remain faithful to the language chosen by Congress," and hold that an activity that "may result in *any* discharge" triggers the certification requirements of section 401(a)(1).

Conclusion

For the foregoing reasons, we conclude that section 401(a)(1) of the CWA requires Alabama Power to obtain water quality certification from the state of Alabama before the Commission can issue a license amendment authorizing Alabama Power to replace the three 33–MW turbine generators at its Martin Dam Project. Because the Commission issued the license amendment to Alabama Power without having such certification, we grant the petition for review and vacate the Commission's orders.

14

River Continuum Concept

Wikipedia

Editor's Introduction: Flora and fauna utilize the energy of water. When, therefore, we remove energy from water through hydropower dams, we necessarily deprive other living beings of that energy. The projects deprive other living beings of their right to the energy of river.

When we take away the kinetic energy though hydropower dams, there occurs a parallel and equal loss of energy available to the other life forms in the river. In other words, production of hydropower is not a 'additional' gain without loss elsewhere. On the contrary, removal of kinetic energy of the river through hydropower projects deprives other life forms of the same energy.

Flow of the river involves dissipation of kinetic energy. This rate and nature of dissipation is affected by width, depth, velocity, flow volume and temperature. Hydropower projects necessarily involve change in these physical characteristics. Thus living organisms have to readjust to the new circumstances.

The availability of energy leads to the establishment of communities of life forms which exploit the available energy. On the other hand, reduction of the same kinetic energy leads to death of such communities. Hydropower projects divert kinetic energy of the stream and, therefore, deprive the river of that energy and impact life forms in the river.

Life forms adjust to stream flow, channel with, depth and gradient, size of organic material, sediment load and temperature regimes. Variation in all these variables occurs in hydropower dams leading to impact on life forms. Our EIAs, however, ignore these impacts. They only examine whether an 'endangered' species is found in the affected stretch. If not, then these impacts on aquatic life are ignored.

Overview

The River Continuum Concept is a model for classifying and describing flowing water. The theory is based on the concept of dynamic equilibrium in which stream forms balance between physical parameters, such as width, depth, velocity, and sediment load, also taking into account biological factors. It offers the introduction to map out pure living communities, but also an explanation for their sequence in individual sections of water. This allows the structure of the river to be more predictable as to the biological properties of the water. The concept was first developed in 1980 by Robin L. Vannote, with fellow researchers at the Stroud Water Research Center.

Background

The River Continuum Concept is based on the idea that a watercourse is an open ecosystem that is in constant interaction with the bank, and moving from source to mouth, constantly changing. Basis for this change in the overall system is due to the gradual change of physical environmental conditions such as the width, depth, water, flow characteristics, temperature, and the complexity of the water. According to the hypothesis, which is based on the physical geomorphological theory, Vannote came up with the hypothesis that structural and functional characteristics of stream communities are selected to conform to the mean state of the physical system. As a river changes from headwaters to the lower reaches, there will be a change in the relationship between the production and consumption (respiration) of the material (P/R ratio).

Living Communities and Food Types

The continuous changes of properties within the river are dependent primarily on the specific composition of the organisms in different sections of the water. Throughout the continuum of the river, the proportion of the four major food types; shredders, collectors, grazers (scrapers), and predators change. With the exception of the predators, all these organisms feed directly from plant material (saprobes).

Shredders. Shredders are organisms that feed off of coarse organic material(CPOM) such as small sections of leaves. They ingest the organic matter along with volunteer organisms (fungi, microorganisms) attached to the source. The preferred size of the CPOM is about one millimeter, therefore shredders must break it up into a finer particulate. In the process of shredding, much of the now finer organic matter is left in the system, making its way further downstream. Some common shredders of North American waters include the mayfly and stone fly larvae.

Collectors. Collector organisms are designated by their use of traps or other adaptive features to filter and catch organic matter. The preferred particle size for collectors lies between 0.5 and 50 micrometers (UPOM = Ultrafine particulate organic matter and FPOM = fine particulate organic matter). This group includes fly larvae, nematodes, and many other animal groups.

Grazers. The grazers (scrapers) feed off of periphyton that accumulates on larger structures such as stones, wood, or large aquatic plants. These include snails, caddisflies (Glossosoma genus), and other organisms.

Because of the differences in the structure of organic matter at different sections in a river, the make up and frequency of these groups in a community vary. In the upper reaches of a river, shredders and collectors make up a large percentage of total macroinvertebrates due to the excess presence of coarse plant matter. In the midreaches of a stream or river, there is an increase in the proportion of grazers due to the presence of periphyton. Shredders only make up a small percentage of the total invertebrates due to the lack of coarse organic matter making its way downstream. In the lower reaches, organic matter has been shredded completely to the level of FPOM or UPOM (Ultra-fine Particulate Organic Matter). Due to the increase in fine particulate organic matter, collectors are the most abundant in the lower reaches, feeding off organic matter and surface film.

The proportion of predators in all sections remains largely constant and only changes in species composition. The reason for the even distribution is that predators are not dependent on the size of the organic matter but on the availability of prey animals in the area.

The Division of the Riverine

The River Continuum Concept assigns different sections of a river into three rough classifications. These classifications apply to all river waters, from small streams to medium-sized and large rivers.

Headwaters (Stream Order 1 to 3). The creek area in the upper reaches or headwaters of a water system is usually very narrow and lined by thick shore vegetation. This prevents the penetration of sunlight, in turn decreasing the production of organic material through photosynthesis in the water. The majority of the organic matter that does makes its way into the system is in the form allochthonous plant material that falls into the river, such as leaves and sticks. In this section, respiration (consumption) out paces production (P/R<1). Here shredders play a major role in breaking down coarse plant material. In this area, the largest diversity of organic material can be expected.

Midreaches (Stream Order 4-6). In the midreaches of a river, river structures such as rocks and trees play an important roll as a supplier of organic material such as periphyton and other autochthonous organic materials. The production to respiration ratio is larger in this section and amounts to P/R>1. The percentage of shredders in this area is less than that of the headwaters, due to lack of coarse plant particulate. Collectors and grazers make up a majority of the macro invertebrate structure in this area, with the predators share remaining unchanged.

Lower Reaches (Stream Order >6). In the lower reaches, there is a large flux in particulate material and also a decrease in production through photosynthesis, due to an increase in water cloudiness and surface film from suspended FPOM. Here, like the headwaters, respiration outpaces production, making the ratio again less than 1 (P/R <1). The living community in these areas are made up of almost exclusively collectors, as well as a small share of predators

Development and Application of the Concept

The first comprehensive presentation of the 1980 concept was part of a two-day conference at Stroud Water Research Center, whose head director was Robin Vannote. It was the result of a multi-year study conducted by the Rockefeller Foundation. The publication of the hypothesis was released later that same year under the title "The River Continuum Concept" in the Canadian Journal of Fisheries and Aquatic Sciences. The concept built on the work of other American Limnologists such as Ruth Patrick, from which the modern riverine ecosystem model has emerged, and Luna Leopold, which deals with the physical changes of water. The essential goal of the concept was to further assess and explain the various communities in the system. Vannote himself described the current situation as follows, "in those days, most people studied a square meter of water to death". Meaning that previous research was always only on small pieces of water and only rarely was the entire river system considered, allowing for the creation of a general model.

After its publication, the River Continuum Concept was quickly the accepted model in the limnology community, becoming a favorite means for describing the communities living in flowing water. Here it broke the classic idea of riverine structure. Previous approaches had their disadvantages because they only described small zones of water and had no consideration for the system in its entirety.

In practice, the River Continuum Concept is used today mainly for environmental assessment of rivers. River studies that assess riverine biological communities and have determined the species composition of an area can then be compared with the ideal species composition from the River Continuum Concept. From there, any variations in species composition may shed light on disturbances that might be occurring to offset the system.

Problems, Limitations, and Modifications

Although the River Continuum Concept is a broadly accepted thoery, it is limited in its applicability. It describes a perfect and even model without taking into account changing riverine disturbances and irregularities. Disturbances such as congestion by dams or natural events such as shore flooding are not included in the model. Various researchers have since expanded the River Continuum Concept to account for such irregularities. For example, J.V. Ward and J.A. Stanford came up with the Serial Discontinuity Concept in 1983, which addresses the impact of geomorphologic disorders such as congestion and integrated inflows. The same authors presented the Hyporheic Corridor concept in 1993, in which the vertical (in depth) and lateral (from shore to shore) structural complexity of the river were connected. The Flood Pulse Concept, developed by W.J. Junk in 1989, further modified by P.B. Bayley in 1990 and K. Tockner in 2000, takes into account the large amount of nutrients and organic material that makes its way into a river from the sediment of surrounding flooded land.

15

The Wild and Scenic Act[1]

Pamela Baldwin

Editor's Introduction: The Wild and Scenic Act was enacted in the United States to preserve free-flowing waters in specified rivers or part thereof. This Act can serve as a model for protecting all other rivers. "Free-flowing" is defined as "existing or flowing in natural condition without impoundment, diversion, straightening, rip-rapping, or other modification of the waterway." Our MOEF, on the other hand, routinely assumes that making barrages for hydropower dams does not affect the flow of the river. The Wild and Scenic Rivers Act creates federal water rights to protected river segments. The federal government has additional power to condemn rights vested under state law if necessary to protect the wild and scenic river.

1. Adapted from "The Wild and Scenic Rivers Act and Federal Water Rights", National Council for Science and the Environment, 1725 K Street, Suite 212 - Washington, DC 20006, 202-530-5810. info@NCSEonline.org, January 2001.

Our Union Government can follow this model to enact a law for protecting the National River Ganga as well as other rivers. We in India are river worshippers. The *kalash,* as embodiment of our rivers, is established in our pooja. Thus our venerable attitude towards rivers stands above that of the U.S. We should have a law stronger than the Wild and Scenic Rivers Act described in this article.

Summary

During the 1960's, support grew for the idea that the development of our nation's rivers needed to be balanced by some means of protecting for future generations certain rivers that possessed outstanding undeveloped qualities. These sentiments culminated in the enactment of the Wild and Scenic Rivers Act of 1968. Rivers may be designated for protection by Congress or, in some cases, be nominated for inclusion by a Governor and approved by the Secretary of the Interior. The Act addresses the protection of the water flows of designated rivers, both expressly and by implication. This report examines the purposes, language, and legislative history of the Act in order to analyze its effects on federal and state water rights.

The Act declares it to be the policy of the United States that certain rivers that possess outstanding values shall be preserved in "free-flowing condition," and that it is the purpose of the Act to implement that policy. The Act contains several paragraphs on water rights, stating that the jurisdiction of the states and United States over waters shall be determined by established principles of law; that any taking of water rights shall entitle the owner to just compensation; that the jurisdiction of the states over waters is unaffected by the Act to the extent that such jurisdiction may be exercised without impairing the purposes of the Act or its administration; and that the Act shall not be construed to alter interstate compacts.

The Act also indicates (albeit by reverse implication) the availability of federal water rights necessary to accomplish the purposes of the Act.

This report discusses federal authority over water, and federal "reserved" and non-reserved water rights. Based on the language of the Act and its legislative history, we conclude that the Act may be interpreted as giving rise to federal water rights. The Act does not specify the quantity of the right. The amount of the federal right is likely to vary from river to river depending on the river's flows, the unappropriated flows in the river at the time of designation, and the values for which the river is being protected. In practice, federal reserved water rights have not always been claimed if alternative means are adequate. Necessary water flows are sometimes secured under state law, through cooperative agreements, and by purchases from willing sellers.

Background

During the 1960's, support grew for the idea that the natural tendency toward development of our nation's rivers needed to be balanced by some means of protecting for future generations certain rivers that possessed outstanding undeveloped qualities. These sentiments culminated in the enactment of the Wild and Scenic Rivers Act of 1968 (WSRA). Rivers may be designated by Congress, or, in some instances, be nominated by a Governor and approved by the Secretary of the Interior. Designation provides certain protections from development and from the adverse effects of water resources projects.

Section I of the Act declares it to be the policy of the United States that certain rivers that possess "outstandingly remarkable scenic, recreational, geologic, fish and wildlife, historic, cultural, or other similar values, shall be preserved in free-flowing condition," and that the established national policy of dam and other construction be complemented by a policy that would preserve other selected rivers or sections "in their free-flowing condition to protect the water quality of such rivers and to fulfill other vital national conservation purposes."

The Act establishes three categories of rivers: wild, scenic, and recreational. A river will be classified as one of these categories depending on its characteristics and values at the time of designation and the desired level of protection. Rivers in the Wild and Scenic River System are managed by various federal agencies.

"River" means a flowing body of water or estuary or section, portion, or tributary thereof, including rivers, streams, creeks, runs, kills, rills, and small lakes.

"Free-flowing," as applied to any river or section of a river, means existing or flowing in natural condition without impoundment, diversion, straightening, rip-rapping, or other modification of the waterway.

The Act provides protection for a designated river or segment by limiting the licensing of dams, reservoirs and other water project works on, or adversely affecting, protected segments. As to the most vital protection of all, the protection of the *flow* of the river, the statute is convoluted, but appears to create federal water rights sufficient to carry out the purposes of the Act. Section 13 states:

(b) The jurisdiction of the States and the United States over waters of any stream included in a national wild, scenic, or recreation river area shall be determined by established principles of law. Under the provisions of this chapter, any taking by the United States of a water right which is vested under either State or Federal law at the time such river is included in the national wild and scenic rivers system shall entitle the owner thereof to just compensation.

(c) Designation of any stream or portion thereof as a national wild, scenic, or recreational river area shall not be construed as a reservation of the waters of such streams for purposes other than those specified in this chapter, or in quantities greater than necessary to accomplish these purposes.

(d) The jurisdiction of the States over waters of any stream included in a national wild, scenic, or recreational river area shall be unaffected by this chapter to the extent that such jurisdiction may be exercised without impairing the purposes of this chapter or its administration.

(e) Nothing contained in this chapter shall be construed to alter, amend, repeal, interpret, modify, or be in conflict with any interstate compact made by any States which contain any portion of the national wild and scenic rivers systems.

The interpretation of this language, its legislative history, and the import for management of wild and scenic rivers will be discussed.

Federal Authority Over Water

Congress derives authority to regulate water from several constitutional sources, among them the commerce power (including the navigation power), the spending power, the war power, the treaty power and the property power. Furthermore, under the "Supremacy Clause," (Art. VI, Clause 2) of the Constitution, when the federal government exercises legitimate authority, the federal law may preempt state law. However, the point at which federal legislation preempts state law is not always clear.

The Supreme Court has recognized the federal power to regulate water under various constitutional powers and resultant statutes. For example, pursuant to the "Commerce Clause" (art. I, § 8, cl.2) in general Congress may regulate water and water use, and, pursuant to the authority to regulate navigation, may even abrogate state sanctioned water rights without paying compensation. Also, the authority for Congress to tax and spend for the general welfare has been said to provide the federal government a range of powers similar to the police power of states to further the public health, safety, and welfare. This clause may provide a source of authority for the imposition of federal directives as to water use in association with the protection of a river.

Considering too that the Act provides for the acquisition of lands by the federal government in the river corridor, another source of constitutional authority for the Wild and Scenic Rivers Act is the "Property Clause" (Art. IV, §3, cl. 2), which authorizes Congress to make "needful rules and regulations" regarding federal property.

Background and Statutory Language

Although Congress has repeatedly deferred to state law in the area of regulation of water use, and a court is likely to be cautious in concluding that a federal water right is created, whether reserved or non-reserved, the power of the federal government to do so cannot be denied. Therefore, the critical factor is whether Congress intended that such rights be created, as indicated either by express language, or by implication from a congressional purpose, reservation, or directive for which water is necessary. A court will derive evidence of that intent from the language of the statute in question, its purposes, and, on points as to which there is any ambiguity, its legislative history.

The purpose of the Wild and Scenic Rivers Act is stated as being to implement the policy set out in section one of the Act - to preserve rivers "in free-flowing condition to protect the water quality of such rivers and to fulfill other vital national conservation purposes." "Free-flowing" is defined as "existing or flowing in natural condition without impoundment, diversion, straightening, rip-rapping, or other modification of the waterway..." Considering the purpose of the Act, it seems likely that Congress intended to create a federal right to some or all of the instream flows of designated rivers or river segments.

This conclusion is reinforced by the express, though negatively stated, reference to the creation of water rights in section 1284(c), that: "designation of any stream or portion thereof as a national wild, scenic or recreational river area shall not be construed as a reservation of the waters of such streams for purposes other than those specified in this chapter, or in quantities greater than necessary to accomplish these purposes."

The words "reserve" and "reservation" also appear in the few Supreme Court cases relating to federal water rights. Although the nature and extent of federal power over water generates perpetual debate, the Supreme Court has held that the federal government may, at the least, "reserve" unappropriated water (water not subject to a right vested under state law) for federal purposes from federal "public domain" lands. This reservation may be express, but typically is inferred from the Congressional purposes in reserving lands for some purpose. The federal right vests and has a priority date as of the date of the reservation, whether or not the water is put to immediate use. Hence, the federal right is senior to all rights vesting after that date. A brief history and discussion of the elements of these rights follows.

"Public domain" lands are federal lands (primarily in the West), that were obtained from a foreign sovereign rather than from a state or individual. These western lands developed an "appropriation" system of water rights that contrasts with the "riparian" rights systems of the eastern states - a system in which water rights depend on ownership of the lands adjacent to the water source. As the original owner of the public domain lands before states were created, the federal government was vested originally with all proprietary and legislative authority predating the authority of states that were subsequently created. Therefore, there could not be any riparian owners for decades because the federal government was the only riparian owner. For this reason, and because water sources in the arid West were scarce and water typically needed to be moved to be used, the new appropriation system of water rights and priorities developed.

The quantity of a WSRA federal water right appears to be the amount necessary to achieve the purposes of the Act; here that would appear to be that amount necessary to preserve the free-flowing condition of the river and to preserve the values for which a river was protected. It is therefore arguable what quantity is sufficient in each instance, and the protected amount may or may not be the fall flow the river. The definition of free-flowing would seem to argue that the full unappropriated flow as of the time of designation (i.e., subject to those existing uses and diversions that do not impair the purposes for which the river is being protected) is protected. On the other hand, by referring to "necessary" water, § 1284(c) may indicate that the amount of the federal right may be less than the full amount of water available. In a river that is subject to heavy Spring flows, for example, the argument might be made that the excess peak water flows could be impounded or diverted upstream as long as sufficient flow was released to the protected segment to maintain its associated fish, wildlife, and scenic values.

Lastly, section 1284(b) states that "under the provisions of this chapter, any taking by the United States of a water right which is vested under either state or federal law at the time such river is included in the national wild and scenic rivers system shall entitle the owner thereof to just compensation." In other words, if any existing water right were ever condemned for federal purposes, compensation would be paid. However, representatives of the Departments of the Interior and Agriculture inform us that no water right has ever been condemned under WSRA.

To summarize, the WSRA appears on its face to protect designated rivers in a free-flowing state by a "reservation" of the waters of such streams, necessary to carry out the purposes of the Act, to affirm existing principles of law as to federal/state authority over water, and to provide compensation for any taking of water rights that were vested under state law.

WSRA Water Rights in Practice

Although federal reserved water rights appear to be available under WSRA, they have not always been claimed. Agency materials indicate that in instances where another underlying federal right (e.g. national forest reserves) exists and appears adequate to provide sufficient water, a WSRA federal right might not be asserted. Similarly, if a right to adequate instream flows is available under state law, the United States has applied for necessary water by that route. Adequate flows may also be obtained under a specific state statute, through cooperative agreements, by filing defensive protests objecting to possibly harmful water right applications by others, or through purchase of necessary water from willing sellers. As noted above, the United States has never condemned water rights for WSRA purposes.

In addition, the individual legislation designating a wild and scenic river may address particular water resource situations.

Discussion and Conclusions

To summarize, the language of the Wild and Scenic Rivers Act itself appears to create federal water rights to flows of protected river segments necessary to carry out the designation purposes. The legislative history fully supports this interpretation. The right appears to be to water not otherwise obligated under state law, but with the additional power in the federal government to condemn rights vested under state law if necessary to accomplish the federal purposes.

Issues remain, however. Because the lands involved in the few Supreme Court cases to date on federal water rights were in western states, the cases involving federal "reserved" rights contained terminology appropriate to public domain lands and "appropriation" water law contexts that prevail in those states. How a court would articulate a federal right in the context of acquired lands or states having a "riparian" water law system cannot be predicted. It is likely, however, that a court would simply articulate the federal right in the manner that best implements Congressional intent, and that the rights associated with federal enclaves in general would be similar to those of federal reservations.

As discussed above, the quantity of the water right is the amount sufficient to carry out the purposes of the Act and no more. Probably, the quantity as to a particular protected river or segment would depend on the existing flows, the values for which the river was being protected, and its classifications.

16

The Wild and Scenic Act in Operation[1]

Interagency Wild and Scenic Rivers Coordinating Council

Editor's Introduction: This paper explains how the Wild and scenic Act is implemented in the U.S. Some of the activities scrutinized are detailed below.

Activities allowed:

- Replacement of existing structures are allowed only if they lead to improvement in free-flow of the river. Abandoned side channel was restored by making a small diversion dam.

---

1. Adapted from Wild & Scenic Rivers Act: Section 7, Appendix A and B, Technical Report of the Interagency Wild and Scenic Rivers Coordinating Council, U.S. Forest Service, Portland, Oregon.

- Construction of a marina for use by employees of a company was allowed because such use was only 2% of the total.
- Restoration of free-flow of the river to remove the effects of past dredging was allowed. Implicitly, dredging is not allowed.

Activities disallowed

- Hydroelectric projects below, above or on stream tributary to the designated river corridor were evaluated for the effects on scenic, recreational, fish and wildlife values denied.
- Tributary of a designated river not allowed to be blocked for irrigation purposes.
- Structures of a proposed highway such as guard rails were redesigned to maintain free-flow of the river. Bridge design was modified to create deeper connection of the river with her flood plain.
- Construction of a boat landing or a dock for barges was denied.

We can repeat these actions for all our rivers.

Applicability of Wild and Scenic Act

The following examples represent typical project proposals that may affect a designated river or a congressionally authorized study river. For some of these examples, whether or not and/or how to apply Section 7 is apparent, for others it may not be as obvious. (Section 7 empowers Federal Government to intervene.) The answers are based on recommendations by river management experts; however, for these, or more complex situations such as mining activities, threatened and endangered species, and hydroelectric development projects, consultation with agency experts and/or legal counsel may be appropriate. While these examples have general application in an agency-identified study river, the land managing agency would achieve the desired outcome through its general authorities and not the Act.

Salmon River Side Channel Restoration, Oregon

Background: The BLM, through development of the RMP, identified an important side channel of the Salmon River for stream habitat restoration. The side channel had been closed 30 years previously by flood control channelization work conducted by the ACOE. During following years, a dike was constructed along the river in order to protect a recreation site and facilities from flooding. Dike construction blocked all flow of Salmon River water to the secondary channel. Restoring flow to the side channel would provide critical off-channel spawning and rearing habitat for anadromous fish. The anadromous fishery of the river was found to be one of the river's ORVs.

Application of Section 7: An interdisciplinary BLM team used the accompanying Section 7 process in preparing an environmental analysis of the proposed project. The Section 7 review analyzed the effects of reopening the secondary channel by breaching the old dike, constructing a diversion structure and diverting flow from the main stem of the Salmon River into the secondary channel for about one-third of a mile before returning to the main stem. The stepwise Section 7 analysis resulted in a careful multi-disciplinary review and design process. The final project satisfied the intent of the Act while meeting stream habitat restoration and flood control goals. Construction of a small head gate was required to maintain the flood control integrity of the dike while allowing control of flows that would mimic natural variations occurring on the main stem. A section 404 permit (under the Clean Water Act administered by the ACOE) was required for construction. The Section 7 analysis satisfied the ACOE, Oregon Department of Fish and Wildlife, and Oregon Department of Water Resources review processes, and the project was approved.

Results: An average of about five to fifteen cubic feet per second (cfs) of water now flows from the main stem of the river through the side channel and adjacent pools, returning to the main stem just one-third mile downstream. The Section 7 analysis provided an alternative that minimized disturbance of the bed and bank at the head of the old channel and reduced sedimentation in the side channel. A surface flow diversion structure with head gate was constructed and screened with natural or natural-appearing materials. Regulation of flow was necessary in order to mimic natural flows, maintain flood protection for the recreation site, and ensure that there will be no adverse impacts to the main stem habitat during extreme low flow conditions. Even at lowest recorded flow conditions (approximately 70 cfs at this location), side channel flows would not exceed 10% of the main channel flows, resulting in negligible effect to the main stem habitat. The project was successful in restoring approximately 2,700 square meters of prime spawning and rearing (particularly over-wintering) habitat for coho salmon, winter steelhead, and cutthroat trout. First year inventories of coho salmon and winter steelhead populations showed that production vastly exceeded expectations. The channel has been highly acclaimed for its habitat restoration, environmental education, research and aesthetic values.

Lower St. Croix River Marina, Minnesota/Wisconsin

Background: Most of the St. Croix Wild and Scenic River in Minnesota and Wisconsin is administered by the NPS, but the lower 25 miles were designated by the Secretary of the Interior and this segment is administered by the States. The Minnesota and Wisconsin Departments of Natural Resources have entered into an agreement with the NPS to form a Lower St. Croix Management Commission so that these three form a management triumvirate.

In 1980, the Control Data Employees Recreational Foundation (Control Data) made application to the ACOE for a permit to construct a marina for its employees in the state-administered segment on the Minnesota side of the river. There already was concern that the Lower St. Croix was being overused for recreational boating, and that the proposed facility would worsen the situation. The state of Minnesota supported construction, but Wisconsin and the NPS opposed it. The Secretary of the Interior wrote to the ACOE requesting that the ACOE withhold a decision pending completion of a river use pattern study being performed under contract for the NPS. The purpose of the study was to determine the appropriate level of river use and development.

Application of Section 7: The project would involve dredge and fill in the bed and on the banks of a navigable waterway included in the National System. It clearly was a water resource development project requiring a Section 10 (of the Rivers and Harbors Act of 1899) permit and a Section 404 (under the Clean Water Act) permit from the ACOE. The NPS felt that it was responsible for reviewing the proposal for possible "direct and adverse effects" under Section 7(a). Control Data claimed the Secretary of the Interior (NPS) did not have this authority for a state-administered river and brought suit. The U.S. District Court for the District of Columbia ruled that the Secretary did have this authority.

Results: A key finding of the study, aside from identifying safety concerns associated with crowded conditions during peak use, was that institutional, industrial and commercial users (which would include Control Data) did not contribute significantly to the volume of traffic (less than 2%) during peak use periods. Therefore, the Secretary lifted his moratorium, and the ACOE issued the permit, but for a smaller marina than had been proposed. The Secretary also ordered that a new policy for future boating facilities be developed with several provisos established. These include: no more large marinas because they contributed significantly to peak use, with the Management Commission to define the distinction between "large" and "small" for Lower St. Croix purposes; establishment of thresholds of safety; and the issuing of permits to individual riparian owners since they also did not contribute significantly to congestion of the water surface or in beach areas.

North Umpqua River Highway Project, Oregon

Background: The Oregon Department of Transportation (ODOT), funded by the federal Public Lands Highway Program, proposed to reconstruct Highway 138, including a segment that parallels part of the North Umpqua Wild and Scenic River. The intent of the proposal was to correct existing pavement and safety deficiencies along the route and to enhance access to the multiple resource activities on adjacent federal lands (managed by the BLM, NPS and USFS). The North Umpqua was added to the National System in recognition of its fishery; water quality and quantity; and recreational, scenic and cultural values. The lower portion of the designated river is administered by the BLM and the upper portion (where a 13.5-mile segment of highway reconstruction was proposed) administered by the USFS.

Application of Section 7: Staff of the Umpqua National Forest utilized the management direction in the recently completed North Umpqua Wild and Scenic River Management Plan to provide input into the state's planning process and as a basis for the resulting Section 7 determination. In addition to identification of the desired future condition for the outstandingly remarkable and other important river values, the North Umpqua Wild and Scenic River Management Plan included guidance for the North Umpqua Highway that is a part of the National Scenic Byway System. Components of the proposed highway project that were evaluated as water resources projects under the "direct and adverse effects" standard included redesign of an access site, widening of a bridge spanning the river, and proposed gabion walls to accommodate roadway width and guardrail.

Results: Early consultation with the ODOT, and the application of the Section 7 procedure by a USFS interdisciplinary team, resulted in significant redesign of the project to protect the river's connection with its floodplain and ORVs. Specifically, the access site redesign eliminated encroachment on bank full channel and lower river terraces in the two- to five-year floodplain. The bridge was widened without modifications or additions to the existing footing and columns (protecting an important anadromous spawning area at the site and immediately downstream), and the gabion wall proposal was eliminated and the site redesigned without additional excavation or embankment construction.

Little Miami River Dock and Concrete Pad Project, Ohio

Background: The Little Miami River was designated in stages (1973 and 1980) by the Secretary of the Interior at the request of the Governor of Ohio. In December 1986, the NPS received a Public Notice from the ACOE announcing that a landowner was requesting "after-the-fact" Sections 10 (of the Rivers and Harbor Act) and 404 (under the Clean Water Act) permits for construction of a boat landing and scenic overlook on a designated segment of the river. The river possessed outstanding scenic, recreational, fish and wildlife, geologic and historic values.

Application of Section 7: On state-administered wild and scenic rivers without adjacent federal ownership, the Secretary of the Interior, through the NPS, is responsible for making Section 7 determinations. (Consultation with the state-administering agencies occurs as needed.) The NPS Midwest Region evaluated this water resources project for "direct and adverse" effects on wild and scenic river values.

Results: After consultation with the Ohio Department of Natural Resources, the NPS determined that the project did have a "direct and adverse effect" on the values for which the Little Miami had been designated. Essentially, the owner's entire riverbank had been converted from a natural condition by covering it with concrete. A letter was sent to the ACOE requesting that the permit not be issued and that the owner be ordered to remove the construction and restore the riverbank.

The ACOE denied the permit and ordered that the compliance measures requested by the NPS be implemented. The landowner refused and in due time the ACOE had the U.S. Justice Department bring suit against the landowner to comply. Eventually, a Consent Order and Judgment was handed down in which the landowner agreed to comply with the terms of the ACOE requirement.

North Fork John Day Dredge Pile Restoration, Oregon

Background: Past dredge mining has severely altered the free-flowing characteristics of the North Fork John Day (North Fork) Wild and Scenic River. The dredge tailings, located over an approximate 11-mile portion of the designated river, have acted to direct the flow of the river against the stream banks, causing stream bank erosion and aggravating a serious bank stability problem. Past dredging also excavated streambed material down to bedrock, completely rearranging the distribution of the natural sediments of the stream. Little recovery has occurred in the 50 years since the site was dredged because of the large amount of tailings and nonerosive characteristics of the dredge piles.

The North Fork is one of the few remaining rivers in the Columbia River Basin that supports wild runs of anadromous Chinook salmon and steelhead. The river also supports populations of bull trout. The purpose of the restoration project is to improve salmonid rearing habitat, water quality, stream bank stability, and floodplain function. The project was based on proven restoration techniques and phased in, beginning with a pilot effort in 1993 that focused on alternative placement of tailings material. Phase 2 expanded the treatment area and extended the time frame to a five-year period.

Application of Section 7: Umatilla National Forest staff applied the accompanying Section 7 process to evaluate the potential for "direct and adverse" effects. The project proposal was coordinated with the Oregon Division of State Lands and the ACOE regarding state/federal permits for fill and removal. In addition, USFS staff consulted with the Oregon Department of Fish and Wildlife, Oregon Parks and Recreation Department (the river is also a State Scenic Waterway), and affected county and tribal governments; the USFS conducted an extensive public outreach as part of the environmental analysis process. The project was designed to remove the artificial impediment to the river's free-flowing characteristics with as little effect to stream bank and riparian vegetation as possible, thereby, allowing the river to reconnect with its floodplain over time.

Results: The results of extensive (and ongoing) monitoring of the first phase of the project indicate that floodplain function has been restored. The river has access to its floodplain and is emulating the natural system as typified in the upstream wilderness segment. The channel morphology is evidencing little change except for widening of the river adjacent to where tailings had been piled. Photo point monitoring shows that most of the pre-project vegetation is still on the site and also that vegetative recovery is occurring on the newly established floodplain. Importantly, the restoration was planned and implemented in a manner to allow the river to reconnect with its floodplain by removing the dredge tailings rather than engineering a solution.

Maurice River Boat Dock and Onshore Mining Development, New Jersey

Background: During the study phase for the Maurice Wild and Scenic River in New Jersey, the Genstar Stone Products Company applied to the ACOE for a permit to build a dock and berthing facility on a section of the river under study. This facility was to service barges for a proposed sand and gravel operation landward of the dock. Construction of the dock and berthing space would require removal of 12,000 cubic yards of sand from the river bottom and banks and construction of 400 feet of bulkhead. In order to move the extracted material from the mine site to the barges, a conveyer system was to be built which would be 60 feet above water level. A 40-foot-high section would extend 85 feet over the river. Ancillary facilities would include roads, a sand processing plant, truck scales, offices, parking and refueling areas. Up to four barge loads would have been moved per day.

Application of Section 7: Because the proposal involved dredge and fill in the river, it was clearly a water resources project and because it would occur in a navigable waterway, an ACOE Section 10 permit was required. The NPS, as the study agency for this river, reviewed the application pursuant to Section 7(b) of the Act.

Results: The effects of the project on the river's free-flowing condition were considered. Some of the ORVs identified by the study were several rare or imperiled plants (specifically the sensitive joint vetch, Parker's pipewort, the sedge *Carex barrattii*, and the boneset *Eupatorium reinosum*); essential bald eagle habitat; and the occurrence of other rare and imperiled birds such as Cooper's hawk, northern harrier and osprey. It was determined that construction activities would occur in some of the same locale as that of the rare and imperiled plants, and that noise from the mining activities and barge traffic would adversely affect the raptor population along the river. There would also be adverse effects on scenic values in the vicinity of the plant, though these values were not judged to be outstandingly remarkable.

Based on this impact assessment, the NPS, through the USFWS, recommended to the ACOE that the permit be denied on the basis of "direct and adverse effects." The New Jersey Department of Environmental Protection found the proposal to be inconsistent with the New Jersey Coastal Zone Management Program. On the basis of these determinations, the ACOE denied the permit.

Sudbury River Aqueduct, Massachusetts

Background: Boston's regional water supply utility, the Massachusetts Water Resources Authority (MWRA), proposed construction of a major new aqueduct in the form of a bedrock tunnel crossing under the Sudbury River in Framingham, Massachusetts. A major tunnel access shaft, "Shaft L," was excavated a few hundred feet from the river. During the construction period, pumps at Shaft L removed the large volume of groundwater that seeped into the tunnel, and the MWRA discharged this water, after treatment, to the Sudbury River. The water treatment facility removed oil and grease, suspended solids, etc., from the water, which was fairly high in water quality. The MWRA proposed construction of a riprap stilling basin on the banks of the river to reduce the velocity of the discharge to a maximum rate of 3200 gallons per minute (or 7.15 cfs). This was the equivalent of almost one-third of the river's natural flow during extreme low flow periods (ten-year drought).

In addition to its free-flowing character, the resources of interest along this portion of the Sudbury included aquatic wildlife habitat (Great Meadows National Wildlife Refuge is one-quarter mile downstream), scenery, recreation, archaeology, and literary (cultural) values (the adjacent oxbow was one of H.D. Thoreau's favorite haunts and features of his writings).

Application of Section 7: The segment of the Sudbury River that was affected by the discharge was, at the time, a congressionally authorized, 5(a), study river and was thus subject to the protections afforded by Section 7(b) of the Act. The river was designated (April 1999) after favorable recommendation by affected towns, the study committee, and the NPS.

The MWRA needed two permits under the federal Clean Water Act: a National Pollutant Discharge Elimination System (NPDES) permit from the Environmental Protection Agency and a Section 404 permit from the ACOE. The NPS, in partnership with the Wild and Scenic River Study Committee, was involved in close consultation with both permitting agencies.

Results: As a result of this process, the following provisions were added to the NPDES and Section 404 permits to ensure that the project would have no direct and adverse effect on the river's values:

The MWRA was not permitted to retain the discharge structure permanently, as originally requested, because of unacceptable direct and adverse impacts on the river's free-flowing character and scenic values.

The proposed stilling basin was moved from a sensitive archaeologic site to a previously disturbed area of riverbank. Riparian wetlands at this site were restored when the basin was removed, enhancing both scenery and riparian habitat.

Discharge velocities were reduced and decentralized to avoid scouring, thereby protecting water quality and aquatic biota. In addition, the MWRA was required to monitor temperature, turbidity, and total suspended solids continuously up and downstream of the discharge point, and to undertake a study of the discharge's effects on benthic macroinvertebrates.

A 70-foot high "mountain" of unvegetated crushed rock that was originally proposed to be created at the site, within view of the river, was eliminated. Instead, rock removed from the tunnel was processed as gravel and sold off site. (While this portion of the project was not subject to Section 7 since it did not require any federal permits or funding, the MWRA chose to mitigate the rock pile's impacts voluntarily.)

PART-V

ECONOMIC VALUE OF FREE-FLOW OF RIVERS

17

Non-use Value of Free Flowing Rivers[1]

John Loomis

Editor's Introduction: Dams are allowed to survive because of the perception that costs of removal exceed the benefits from retaining. This perception arises because some of the benefits of dam removal are ignored—especially the passive or non-use values.

The fact that a species is endangered means that economic value cannot be created from its 'use'. If the number of lions is small, it will not be possible for a large number of people to derive value from watching them. But people can be happy knowing that the lion lives even though they cannot see it. This happiness is 'non-use' value. This benefit accrues to a person even though the person does not ever use the resource.

1. Adapted from "Importance of Including Use and Passive Use Values of River and Lake Restoration," Universities Council on Water Resources Journal of Contemporary Water Research & Education, Issue 134, Pages 408, July 2006.

People secure benefit from knowing that a species exists for future generations.

This concept applies strongly to our rivers. Our people may never visit Haridwar and take bath in the Ganga. But they have great veneration for this river. A Keralite would derive some satisfaction knowing that the Ganga flows free. This benefit, utility or economic value is comparable to the use value, say, that is derived from electricity through reading of a book. Both give utility or satisfaction. The problem is that the satisfaction obtained from knowledge of free flow of the Ganga is ignored while the satisfaction obtained from the consumption of electricity is shown in calculations. This mistake in accounting leads to the building of dams for generation of electricity.

The paper gives case studies that show that one-half of the benefits of free-flowing rivers accrues in the form of such non-use values. We, therefore, reduce the benefits of free-flowing rivers by one half when we ignore these values. Decommissioning of certain dams in the United States became economically beneficial once the non-use values of free flowing rivers were included. The inclusion of non-use values in Cost-Benefit Analysis of dams proposed on Indian Rivers can likewise render the projects uneconomic.

The non-use value is assessed by providing a person with two scenarios of existing- and improved river conditions and then asking how much they would be willing to pay for securing the improved condition.

The objective of producing hydropower is to secure welfare of the people by providing electricity. The same welfare is also secured by improvement in the condition of our rivers, wildlife, etc. Unfortunately, our planners assess the benefits from hydropower but ignore the deduction of the same welfare due to destruction of our rivers. It is possible that the loss of welfare from destruction of free flow of a river may be more than the increase from generation of electricity. In such case we are creating projects which have a negative impact on the people.

Restoration is an Economically Important Activity

The restoration of rivers and related riparian areas is now a billion dollar a year business, with at least $15 billion spent since 1990 (Bernhardt et al. 2005). This restoration is taking place coast to coast, from the Penobscot River in Maine to the Elwha River in Washington. Restoration brings hope and optimism to conservationists that some of the past injuries that have arisen from our overzealous development of rivers can be at least partially rectified.

But as river restoration grows into a billion dollar a year effort, certainly there will be individuals that will ask whether the benefits of such efforts are worth the costs. Not every restoration effort such as the Everglades will have the President of the United States' brother to advocate for it. Some proposed large-scale restoration efforts such as the lower Snake River dam removals have yet to be approved, in part because of the perception by politically powerful interests that the costs outweigh the benefits. This paper will show that this perception is in part due to omission of important passive use values of river and salmon restoration such as existence values. These non-use or passive use values of river restoration are critical to include when dealing with restoration of riverine habitat for threatened and endangered (T&E) species. The fact that a species population is so low that it is listed under the Endangered Species Act should suggest that the economic justification for restoration of its habitat will not come from commercial and recreational use values; the current and near future populations are just too low. The near term value to society lies elsewhere, in the passive use value component of total economic value.

What is Total Economic Value of Restoration?

The Total Economic Value (TEV) associated with restoration is made up of the obvious on-site use value, as well as the not so obvious (at least to some) off-site passive use values. The on-site use values of river restoration include a wide variety of ecosystem services such as recreation, fish habitat, water quality, stormwater management and aesthetics. However, restoration also provides widespread benefits to people who obtain satisfaction or utility from knowing that native species exist in their natural habitat (i.e., existence value) or from knowing that restoration today provides native species and their natural habitats to future generations (i.e. a bequest value).

These existence and bequest values have been termed passive use values since they were upheld by the U.S. Court of Appeals for use in natural resource damage assessment. In ruling against the U.S. Department of Interior's (DOI) damage regulations proposal to only allow either use or non-use values to be counted, the U.S. Court of Appeals noted: "Option and existence values may represent 'passive' use, but they nonetheless reflect utility derived by humans from a resource and thus, prima facie, ought to be included in a damage assessment." (U.S. Court of Appeals 1989: 67). In response to this court ruling, DOI agencies include use and passive use values in their natural resource damage assessment (Ward and Duffield 1992, USDOI 1994).

My thesis is that, since passive use values are appropriate for the government to collect when damages occur, passive use values are appropriate to include when estimating the benefits of river restoration as well.

Techniques for Estimating the Use and Passive Use Values of Restoration Use Values of River Restoration

To estimate use values of river restoration, economists often rely upon actual market behavior to detect how visitors or homeowners value river restoration. Visitors reveal their greater demand and benefits for improved rivers by the increased number of trips they take to restored streams and rivers as compared to degraded ones. The Travel Cost Method (Loomis and Walsh 1997) can be used to estimate the demand curve for restored rivers and allows for the calculation of the visitor's additional net willingness to pay to visit these restored rivers, as compared to degraded ones. For rivers running through residential areas, house price differentials reflect what homeowners will pay for living by a restored or natural stream as compared to a degraded one. This statistical analysis of house price differentials is called the Hedonic Property Method.

Passive Use Values of River Restoration

Existence and bequest values do not leave obvious behavioral trails and so economists have developed constructed or simulated markets to allow people to state what they would pay to know that a restored river exists with native fish. The two types of stated preference approaches are Contingent Valuation Method (CVM) (Mitchell and Carson 1989) and conjoint/choice experiments (Louviere et al. 2000). Both methods involve providing households with a comparison of existing river conditions and improved river conditions and then ask whether they would pay a given increase in cost that varies across households. The varying costs and the response to them allow for tracing out a demandlike relationship for restoration (i.e., the higher the cost, the fewer people would pay). CVM estimates a value for the entire restoration improvement program (Loomis 1996), while conjoint allows for the valuation of each individual ecosystem service provided by the restoration.

While reliance on what people say they would pay has been controversial (Portney 1994), the method has shown to be reliable in test-retest studies (Loomis 1989a 1990, Reiling et al. 1990, Carson et al. 1997). Past comparisons with actual cash have shown that CVM derived values may overstate true WTP. However, a blue ribbon panel appointed by the National Oceanic and Atmospheric Admisnistration (NOAA) that was chaired by two Nobel Laureates concluded that carefully constructed CVM studies are believed to yield reliable enough estimates of existence or non-use value to be a useful starting point for judicial and administrative deliberations (Arrow, et al., 1993).

Empirical Examples of Use Values of River Restoration

The hedonic property method has been frequently employed to estimate the value of river restoration. One of the first applications was by Streiner and Loomis (1995), which showed that houses in northern California along streams that were restored sold for 11 to 12 percent more than houses along unrestored streams. Research in Arizona by Colby and Wishart (2002) and Colby et. al. (2005) suggest that riparian areas have a significant positive influence on property values. Netusil (2005) found that publicly owned streams had a positive and significant influence on property values in Oregon. Houses located at greater distances from lakes had lower values in Connecticut, which suggests the importance of water resources to house prices there (Acharya and Bennett 2001). Water quality was found to have a significant effect on house prices at several lakes in Maine (Boyle and Taylor 2001).

Empirical Examples of the Importance of Including Passive Use Values in River Restoration

Restoration of free-flowing rivers and recovery of native species often has existence values that are received by households all across the entire nation (Loomis 2000). Previous studies have shown that existence values make up at least half the benefits of improving water resources (Fisher and Raucher 1984, Sanders et al. 1990), the majority of benefits for many rare bird species (Loomis 1989b), along with T&E species. It is important to include these passive use or non-use benefits when calculating the benefits of restoration. The empirical importance of doing so is illustrated with two case studies. The first one involves removal of the Elwha dam and Glines dam to open up 70 miles of the Elwha River to native salmon, Loomis.

Elwha Case Study

The removal of two dams from the Elwha River near Olympic National Park in Washington is an expensive proposition with costs of nearly $250 million. It will take decades before significant increases in harvestable fish return to support appreciable commercial and recreational fishing. But the restoration of the river and return of the natural migration of the salmon is expected to occur within the first decade. Thus, most of the near term benefits to Washington residents are existence values or passive use values, not use values. In order to estimate the passive use values associated with the dam removal and river/salmon restoration, a CVM survey was conducted in which Washington households were asked about their willingness to pay for dam removal and salmon recovery. The willingness to pay question was framed as a voter referendum question, asking whether they would vote in favor of dam removal and salmon restoration at a specific increase in cost. This cost, ($X), varied from $3 to $190 across the sample. The increase in salmon populations with dam removal versus fish ladders was illustrated with a bar chart. The wording of the willingness to pay question was:

- If a majority are not willing to pay, the dams remain and fish populations are as shown for "Dams."
- If a majority agree to pay the cost, the dams would be removed, river restored and fish populations would increase as shown in "Dam Removal."
- If an increase in your federal taxes for the next 10 years cost your household $X each year would you vote in favor? YES/ NO.

The survey response rate was 68 percent for Washington residents, and their average WTP was $73 (with a 90 percent confidence interval of $60-$99). This translates into about $94 million in passive use values to Washington households each year. Including these passive use value results in positive net benefits (benefits in excess of cost) for dam removal.

Lower Snake River Dam Removal

The U.S. Army Corps of Engineers conducted a study on whether dam removal was a reasonable and prudent alternative for salmon recovery on the lower Snake River. This 140-mile stretch of the Snake River has four dams, which essentially convert this 140-miles into four slack water reservoir pools. This slack water greatly slows out-migration of smolts to the

Columbia River and to the Pacific Ocean. The net result is higher than natural mortality for the salmon, and three of the Snake River salmon species are listed as either threatened or endangered.

Unfortunately, the official benefit-cost guidelines of the COE, the U.S. Water Resources Council Principles and Guidelines, were last updated in 1983. This was several years before the measurement of passive use value was routine or mandated by the U.S. Court of Appeals for Interior agencies. Including only the recreation use and commercial fishing values for salmon and steelhead populations results in dam removal having a large negative net benefit (-$267 million) relative to artificially transporting fish around the dams (+$13.5 million)—see U.S. Army Corps of Engineers (2002).

The COE originally planned to include a nonuse value question in its economic analysis, but intervention by then Washington Senator Slade Gorton (who supported dam retention) resulted in the question being removed from the survey. Calculation of the non-use value for salmon and free flowing rivers was done using benefit transfer from existing non-use valuation studies. Using a variety of benefit transfer protocols, the passive use value of salmon was estimated for the dam removal alternative at between $22.8 and $310.5 million (U.S. Army Corps of Engineers 2002: 42). The passive use value of restoring the free flowing river was estimated at $420 million (U.S. Army Corps of Engineers 2002: 42).

As shown by including the passive use values of the free-flowing river and salmon restoration, this would make dam removal economically efficient, which yields the highest net benefits. The omission of the passive use values may have contributed to the Corps of Engineers decision to keep the dams in place.

Conclusions

As the two case studies illustrate, calculating the total economic values of restoration including passive use values is necessary so as to not understate the benefits of river restoration. The passive use or non use values often make up a majority of the benefits, and their omission can often lead to the impression that the restoration is uneconomic. As these two case studies indicate, the inclusion of passive use values demonstrates that restoration was economically efficient, with the benefits exceeding costs. While economics should not be the sole determinant of whether to restore an area or not, as restoration projects expand in frequency and scale, some prioritization of restoration projects becomes inevitable. In sorting through restoration projects that compete for scarce funding, having information about the use and passive use values of the restoration project can aid decision-makers in selecting restoration projects that provide the greatest benefits to society as a whole.

References

Archarya, G. and L.L. Bennett. 2001. Valuing Open Space and Land Use Patterns in Urban Watersheds. *The Journal of Real Estate Finance and Economics* 22(2): 221-238.

Bernhardt, E. and 12 others. 2005. Synthesizing U.S. River Restoration Efforts. *Science* 308: 636-637.

Boyle, K. and L. Taylor. 2001. Does the Measurement of Property and Structural Characteristics Affect Estimated Implicit Prices for Environmental Amenities in a Hedonic Model? *The Journal of Real Estate Finance and Economics* 22(2): 319-338.

Carson, R., M. Hanemann, R. Kopp, J. Krosnick, B. Mitchell, S. Presseor, P. Ruud, V. K. Smith, M. Conaway, and K. Martin. 1997. Temporal Reliability of Estimates from Contingent Valuation. *Land Economics* 73(2): 151-163.

Colby, B. and S. Wishart. 2002. Quantifying the Influence of Desert Riparian Areas on Residential Property Values. *The Appraisal Journal* 70 (3): 304-308.

Colby, B., R. Bark and D. Osgood. 2005. Valuing Restoration of Water Dependent Urban Amenities. *Proceedings of the UCOWR / NIWR Annual Conference*. Portland Maine.

Fisher, A. and R. Raucher. 1984. Intrinsic Benefits of Improved Water Quality: Conceptual and Empirical Perspectives. In V. K. Smith, ed. *Advances in Applied Micro-Economics*. JAI Press Inc, Greenwich, CT.

Loomis, J. 1989a. Test Re-test Reliability of the Contingent Valuation Method: A Comparison of General Population and Visitor Responses. *American Journal of Agricultural Economics* 71(1): 76-84.

Loomis, J. 1989b. Quantifying the Economic Value of Public Trust Resources Using the Contingent Valuation Method: A Case Study of the Mono Lake Decision. *Transactions of the 54th North American Wildlife and Natural Resources Conference.* Wildlife Management Institute, Washington DC.

Loomis, J. 1990. Comparative Reliability of the Dichotomous Choice and Open Ended Contingent Valuation Techniques. *Journal of Environmental Economics and Management,* 18(1): 78-85.

Loomis, J. 1996. Measuring the Economic Benefits of Removing Dams and Restoring the Elwha River: Results of a Contingent Valuation Survey. *Water Resources Research* 32(2): 441-447.

Loomis, J. 2000. Vertically Summing Public Good Demand Curves. *Land Economics* 76(2): 312-321.

Loomis, J. and R. Walsh. 1997. *Recreation Economic Decisions, 2nd Edition.* Venture Publishing, State College, PA.

Louviere, J., D. Hensher, and J. Swait. 2000. *Stated Choice Methods: Analysis and Application.* Cambridge University Press. New York, NY.

Mitchell, R. and R. Carson. 1989. *Using Surveys to Value Public Goods: The Contingent Valuation Method.* Resources for the Future, Washington DC.

Netusil, N. R. 2005. Does Ownership Matter? Examining the Relationship Between Property Values and Privately and Publicly Owned Open Spaces, Streams and Wetlands. Reed College, Department of Economics, Working Paper.

Portney, P. 1994. The Contingent Valuation Debate: Why Economists Should Care. *Journal of Economic Perspectives* 8(4): 3-18.

Reiling, S., K. Boyle, M. Phillips, and M. Anderson. 1990. Temporal Reliability of Contingent Values. *Land Economics* 66(2): 128-134.

Sanders, L., R. Walsh, and J. Loomis. 1990. Toward Empirical Estimation of the Total Value of Protecting Rivers. *Water Resources Research* 26(7): 1345-1358.

Streiner, C. F. and J. Loomis. 1995. Estimating the Benefits of Urban Stream Restoration Using the Hedonic Price Method. *Rivers* 5 (4): 267-78.

U.S. Army Corps of Engineers. 2002. *Improving Salmon Passage.* Final Lower Snake River Juvenile Salmon Migration Feasibility Report/Environmental Impact Statement. Walla Walla District, Walla Walla WA.

UCOWR *Journal of Contemporary Water Research and Education* United States District Court of Appeals for the District of Columbia Circuit. 1989. State of Ohio vs U.S. Department of Interior, et al. Cases # 86-1529 and # 86-1575. July 14, 1989.

U.S. Department of Interior (USDOI). 1994. Natural Resource Damage Assessments; Final Rule. 43 CFR Part 11. *Federal Register* 59(58): 14262-14288. Ward, K. and J. Duffield. 1992. *Natural Resource Damages: Law and Economics*. John Wiley and Sons. New York, NY.

18

Economic Value of Rivers[1]

New Jersey Department of Environmental Protection

Editor's Introduction: The reader would be familiar with the terms 'capital' and 'revenue' expenditures. Capital expenditures are those which provide services for an extended period such as planting of an apple tree. Revenue expenditures, on the other hand, are those that provide services for one event such a purchase of a box of apples. A similar distinction applies to natural resources. A forest provides carbon sequestration for its life time while it provides timber as a single event. Capital expenditures are generally considered 'good' while revenue expenditures are considered, relatively speaking, 'bad'. If we view a project as a source of long-term benefit, we are more

---

1. Adapted from "Valuing New Jersey's Natural Capital: An Assessment of the Economic Value of the State's Natural Resources," April 2007.

likely to invest in it than if we view it as raw material to be used up in the near future.

This distinction also applies to our rivers.

Rivers provide certain long term services such as sediment renourishment to our coasts, biodiversity, aesthetic value, etc. They also provide 'short term' benefits like hydropower. These are 'short term' since dams have a life period of about 50-70 years against perpetual value of free-flow of rivers.

This paper makes an assessment of the capital value of forests of New Jersey. There is much pressure to use forest land for residential or commercial purposes. The case made for development projects usually includes quantitative projections of claimed economic benefits, such as jobs, tax revenues, etc. But the long run environmental benefits from the forests are either not recognized or mentioned only qualitatively. This makes it impossible to compare environmental costs with other asserted costs and benefits.

An assessment of the value of various long term services of natural resources is presented in this paper. This paper also mentions that there is a 'scarcity value' of natural resources. The value of a particular resource rises as it becomes scarce. The use of our rivers for hydropower leaves less free flow and thus enhances its value. Our Forest Conservation Act explicitly provides that a Cost-Benefit Analysis of the activity proposed on the diverted forest land will be presented in the application for permission for diversion of forest lands.

The Cost-Benefit Analysis presented by hydropower projects does not present the value of long term services lost by preventing free flow of rivers. In this way the costs of hydropower are artificially reduced. Thus, when we change the flow of river we should account for the loss of these services. But, in absence of clear estimates, these lost services are mentioned, if at all, qualitatively and loss to the economy from disabling these services is not accounted for. Extracts are produced from the New Jersey study to show the high economic value of services provided by river systems. The study shows that freshwater wetlands have a value of $11,802/acre/year and riparian buffer has a value of $ 3,500/acre/year which is much higher than forest land at $1,714. This value of water resources should be included in economic assessments of water diversion. In particular, hydropower plants often lead to tracts of river becoming dry. This involves a huge loss of economic value that is not accounted.

In 2004, the New Jersey Department of Environmental Protection partnered with the Geraldine R. Dodge and William Penn Foundations to undertake an important project to assess the economic value of New Jersey's natural resources.

The Concept of Natural Capital

Before we can discuss "natural capital", we need to understand the concept of "capital" in general and the related concept of "assets". In economics, assets are entities that possess "exchange" (i.e., market) value and that form part of the wealth or property of their owner (Pearce 1992). Capital assets are assets that generate a flow of economic benefits over an extended period. In contrast, the value of operating assets such as gasoline, office supplies, and food is usually used up relatively quickly.

Perhaps the most familiar types of capital assets are physical capital such as machinery, buildings, equipment, etc., and financial capital; other types of capital recognized by economists include human capital (e.g., a population's numbers, skills, training, etc.) and social or cultural capital (e.g., the ability to own property and enforce contracts and the other institutions that make private economic activity possible). In each case, the use of the term "capital" emphasizes the fact that the assets in question provide value over an extended period.

In recent years, many economists have begun using the term natural capital to call attention to the fact that various naturally-occurring entities also provide economic value over an extended period, a period that for some assets is essentially perpetual on any meaningful human time scale. The term "natural capital" differs from the older term "natural resources" in that the latter views nature as essentially a source of raw materials which lack value until they are extracted from their natural environment and put to use. "Natural capital" also differs from "natural environment" in emphasizing nature's role as an active source of economic value.

These distinctions are of great practical importance. If we view something as a long-term source of benefit, we are more likely to invest in maintaining its productive capacity than if we view it as raw material to be used up in the near future. For example, if a forest is seen only as a source of short-term profits on timber sales, there is no particular reason to delay harvesting the resource and reaping the benefits. However, to the extent that the forest is seen as a capital asset, the owner has an incentive to limit the amount of logging in some way to preserve the forest's long-term profitability. This incentive is increased if the forest is seen as an asset that provides things of value in addition to wood, e.g., recreational opportunities.

Many of the benefits provided by natural capital come from ecological systems ("ecosystems"); an ecosystem is a dynamic complex of plant, animal, and micro-organism communities and their nonliving environment, all interacting as a functional unit (UNEP 2001-2005). The benefits provided by natural capital include both goods and services:

- "Ecosystem goods (such as food) and services (such as waste assimilation) represent the benefits human populations derive, directly or indirectly, from ecosystem functions" (Costanza et al. 1997).
- "Ecosystem services are the conditions and processes through which natural ecosystems, and the species that make them up, sustain and fulfill human life. They maintain biodiversity and the production of ecosystem goods, such as seafood, forage timber, biomass fuels, natural fiber, and many pharmaceuticals, industrial products, and their precursors" (Daily 1997). Examples of

ecosystem services ("ecoservices") include temporary storage of floodwaters by wetlands, long-term storage of climate-altering greenhouse gases in forests, dilution and assimilation of wastes by rivers, and numerous others.

- Goods are physical commodities that can be weighed, packaged, and transported. Some classification systems treat nature's provision of goods as a type of service ("provisioning" services); for convenience this study treats goods separately. Although goods can come from both ecosystems (e.g., timber) and abiotic (non-living) sources (e.g., mineral deposits), for convenience all goods deriving directly from natural sources are referred to in this report as "natural goods", "ecosystem goods", or "ecogoods".

The Natural Capital Project

As the most densely populated state in the U.S., New Jersey is under more or less constant pressure to convert undeveloped land to residential, commercial, and other uses; the potential for such conversion is one of the top environmental issues for the state's residents and businesses. The case made for development projects usually includes quantitative projections of claimed economic benefits, such as jobs, property tax revenues, etc. The arguments made against development increasingly include quantitative projections of claimed economic costs, such as the cost of new schools, new or expanded highways, etc.

While projected environmental costs are often part of the case made against land conversion, this type of cost is often expressed in qualitative terms; where it is quantified, the figures cited (e.g., acres of wetlands lost) are usually not expressed in monetary terms. This makes it essentially impossible to quantitatively compare environmental costs with other asserted costs and benefits. Some would say that this inability is for the best, since it protects environmental assets from being lost due to hasty or otherwise deficient benefit-cost analyses. Others believe that the natural environment can most effectively be protected if its value as natural capital— expressed in monetary terms—is widely understood.

Fig. 18.1. River Bank.

Components of Natural Capital

In economics and finance, the value of a capital asset is determined by the value of the future benefits which the asset is expected to provide; in effect, the benefits represent income in the accounting sense, while the asset values make up part of the balance sheet. Natural capital provides two main types of direct benefit: services (such as removal of suspended solids by wetlands) and goods or commodities (such as timber). In addition, natural capital can be divided into biotic (living) systems such

Table 18.1

Types of Natural Capital and Direct Economic Benefits

Natural Assets	*Value from services*	*Value from goods*
Biotic systems, e.g., wetlands.	Ecosystem services, e.g., sediment removal.	Ecosystem goods, e.g., fish.
Abiotic systems.	Dilution of air pollutants.	Provision of groundwater.

as forests and wetlands and abiotic systems such as underground aquifers and mineral deposits. These distinctions give rise to the following classification scheme:

In addition to these types of direct benefit, the natural environment also provides the essential setting for the production of what might be called indirect benefits, such as those generated by ecotourism. The benefits of ecotourism differ from the benefits of ecosystem services and natural goods because they derive from the fact that visitors to natural sites spend money in connection with their visits; and those expenditures in turn generate further economic activity as the dollars involved are re-spent. If visitors to natural sites spent a bare minimum on their activities (e.g., getting to the site and back but nothing else), the benefits of ecotourism would decline substantially, but the ecosystems involved would be unchanged, as would the value of the natural goods and ecoservices they provide. For these and other reasons, assessing the economic effects of ecotourism requires different methods from those used for ecosystem services and natural goods.

Approach for Ecoservices and Ecogoods

The value of a capital asset is determined by the value of the services and goods which that asset provides over time. This simple statement reflects a number of important principles and assumptions, as described in the following paragraphs.

Level of analysis. Technically, each locality is unique, but to conduct any kind of analysis at that level of specificity is not realistic at present. Since the goods and services provided vary by ecosystem, the two studies presented in this report used that level of aggregation, e.g., all New Jersey forests, all New Jersey wetlands, etc. Where possible, important distinctions are made at the ecosystem level, e.g., between forested and unforested wetlands.

Natural capital metric. To estimate the dollar value of New Jersey's natural capital, we need to know how much natural capital the state has in a physical sense. The generally used metric for this, and the metric used in the present studies, is acreage. As with the level of analysis, this metric treats all acres of a given ecosystem type as fungible, even though each acre may be unique in some relevant sense.

Treatment of time. Capital assets produce value over an extended period, and each year's values must therefore be combined to produce a single "present" value for the asset. In keeping with the standard practice in economics and other fields, this is accomplished by mathematically "discounting" the values of goods and services provided in future years. In essence, the annual future benefits stream is assumed to be constant, and that constant value is discounted at 3% per year in perpetuity to obtain the present value of the natural asset.

Goods vs. services. The need to avoid double-counting of benefits is always a consideration in studies such as the present ones, and the researchers involved in this project have taken care to avoid such double-counting.

Results by Ecosystem

Table 18.2 summarizes the results by ecosystem; annual values ($MM and $/acre) and present values ($Bn and $/acre) are given. The ecosystems are listed in order by the total value of goods plus services.

For both goods and services separately and for the two combined, the figures in Table 18.2 clearly demonstrate a wide range of both values per acre and total values, spanning two orders of magnitude. Every system except beaches/dunes, barren land, and paved urban land provides both goods and services. Beaches/dunes have by far the highest dollar value per acre due to the extremely high value that many people place on the services provided by this ecosystem.

Table 18.2 also shows that on an overall basis, New Jersey's ecosystems are far more valuable as providers of services than

Table 18.2 : Value of New Jersey's Natural Capital (excluding ecotourism) (2004 $)

Ecosystem	Area (acres)	Natural Goods				Natural Services				Natural Goods & Services			
		$/MM/yr	$/ac/yr	PV $Bn	PV $/ac	$MM/yr	$/ac/yr	PV $Bn	PV $/ac	$MM/yr	$/ac/yr	PV $Bn	PV $/ac
Freshwater wetland [1]	814,479	$191	$234	$6.4	$7,801	$9,422	$11,568	$314.1	$385,593	$9,612	$11,802	$320.4	$393,394
Marine[2]	755,535	850	1,125	28.3	37,512	5,700	7,544	190.0	251,475	6,550	8,670	218.3	288,987
Farmland**[3]	673,464	3,760	5,583	125.3	186,095	483	717	16.1	23,887	4,242	6,229	141.4	209,982
Forest land***	1,465,668	349	238	11.6	7,934	2,163	1,476	72.1	49,201	2,512	1,714	83.7	57,136
Saltwater wetland	190,520	26	139	0.9	4,617	1,168	6,131	38.9	204,355	1,194	6,269	39.8	208,973
Barren land	51,796	587	11,337	19.6	377,893	0	0	0.0	0	587	11,337	19.6	377,893
Urban[4]	1,483,496	20	13	0.7	450	419	283	14.0	9,420	439	296	14.6	9,869
Beach/dune	7,837	0	0	0.0	0	330	42,149	11.0	1,404,969	330	42,149	11.0	1,404,969
Open fresh water	86,232	79	921	2.6	30,689	66	765	2.2	25,510	145	1,686	4.8	56,208
Riparian buffer	15,146	2	118	0.1	3,934	51	3,382	1.7	112,747	53	3,500	1.8	116,681
Total or Avg.	5,544,173	$5,864	$1,058	$195.5	$35,259	$19,802	$3,572	$660.1	$119,059	$25,667	$4,630	$855.6	$154,317
1. Freshwater wetlands													
Forested	633,380	154	244	5.1	8,122	7,327	11,568	244.2	385,593	7,481	11,811	249.4	393,715
Unforested	181,099	36	200	1.2	6,679	2,095	11,568	69.8	385,593	2,131	11,768	71.0	392,272
2. Marine													
Estuary/tidal bay	455,700	513	1,125	17.1	37,505	5,310	11,653	177.0	388,448	5,823	12,779	194.1	425,953
Coastal shelf	299,835	338	1,126	11.3	37,524	389	1,299	13.0	43,297	727	2,425	24.2	80,820
3. Farmland **													
Cropland	546,261	3,291	6,025	109.7	200,828	473	866	15.8	28,855	3,764	6,890	125.5	229,683
Pasture/grassland	127,203	469	3,685	15.6	122,827	10	77	0.3	2,551	478	3,761	15.9	125,379
4. Urban													
Urban (impervious)	1,313,946	0	0	0.0	0	0	0	0.0	0	0	0	0.0	0
Urban green space	169,550	20	118	0.7	3,934	419	2,473	14.0	82,420	439	2,591	14.6	86,354

* middle estimates, including consumer surplus. ** ecosystem service values for farmland have been revised. *** includes wooded farmland.

as sources of harvestable goods, a fact that has important implications for land use and environmental protection.

Finally, Table 18.2 demonstrates the high value of New Jersey's natural capital: $25.7 billion/year for goods and services combined (just over $4,600/acre/year) and $856 billion in present value ($154,000/acre). Freshwater wetlands and marine ecosystems have the highest total values.

Based on this analysis, the total value of New Jersey's natural capital appears to be about $681 billion, plus or minus $300 billion. A range of this magnitude is not surprising given the complexity of the ecosystems being analyzed and the uncertainties in each of the many component estimates that make up these grand totals.

Conservatism of the Estimates

The results summarized above have to be regarded as initial estimates of economic value rather than as definitive conclusions. In part, this is due to the fact that those results had to leave out a number of sources of value, including (but not limited to) the following:

1. *Limited coverage.* This is perhaps the most important issue. Some ecosystems and ecoservices have not been very well studied, and some have not been studied at all. For example, the results do not reflect the value of the genetic data contained in New Jersey's natural capital, i.e., its plant and animal life. More comprehensive coverage would almost certainly increase the values shown in this report, since no valuation studies to date have reported values of less than zero.
2. *Scarcity value.* The valuations probably underestimate shifts in the relevant demand curves as the supply of natural capital declines due to continued conversion of undeveloped land to other uses. Such shifts would in all probability result in an increase in society's willingness to pay for the natural capital that remains. If New Jersey's

ecosystems are now smaller than assumed here, their value is therefore probably underestimated in this study. Such reductions appear likely as land conversion and development proceed; climate change may also adversely affect New Jersey's ecosystems, although the precise impacts are harder to predict.

3. *Omitted value components*. Because the value transfer method used for ecoservices is based on average rather than marginal cost, it cannot provide estimates of consumer surplus. However, this means that ecoservice valuations based on averages are more likely to underestimate total ecoservice value.
4. *Externalities.* Distortions in the market prices used to estimate ecoservice values are unavoidably carried through the analysis. These prices do not reflect environmental externalities and are therefore again likely to be underestimates of "true" values.
5. *Secondary effects*. The values reported only reflect "direct" effects, but "secondary" effects may also be important for some of the goods and services studied. When costs are incurred to produce and distribute natural goods, or when costs are avoided because natural ecoservices eliminate the need for investment in artificial substitutes, at least some of the expenditures made (or the expenditures made with funds saved) stimulate "secondary" economic activity, e.g., as when farmers purchase supplies or equipment or when employees of mining companies spend their wages on goods and services.
6. *Existence value*. The results do not fully reflect what economists refer to as "existence value". It is well known that people value the existence of certain ecosystems, even if they never plan to use or benefit from them in any direct way. Estimates of existence value in the peer-reviewed literature are rare, and fully including this "service" would again increase the total values.

All of these factors lead to under-estimates of value, and there are relatively few factors that would cause over-estimates.

Other Ecosystem Services

The results presented in this report do not address all of the ecosystem services provided by New Jersey's natural capital. These deliberate omissions reflect various factors, including absence of peer-reviewed studies, unavailability of data, lack of accepted analytic techniques, etc. However, these omissions necessarily lead to an understatement of the total value of New Jersey's natural capital. To illustrate the potential magnitude of this understatement, this section presents analyses of several forest ecoservices for which peer-reviewed studies were not available.

As this report was being finalized, the Intergovernmental Panel on Climate Change (IPCC) issued its long-awaited Fourth Assessment Report on the causes and consequences of global climate change (see www.ipcc.ch). The report contains the IPCC's clearest warnings to date on the adverse impacts caused by global warming and the ways in which human emissions of greenhouse gases (GHG's)—especially carbon dioxide—are worsening those impacts.

The results of the current study reflect the estimated value of the sequestration (removal from the atmosphere) by New Jersey's forests of carbon dioxide. For technical reasons, the current study does not discuss in detail the value of previous forest storage of carbon dioxide. Given the growing recognition of the importance of slowing the growth of GHG emissions, Crude estimates of the value of the carbon storage service provided by New Jersey's forests could range from $3.5 to $10.4 billion in present value terms. Because they have not been published in peer-reviewed journals, these amounts are not included in the totals presented in this report, which again underscores the conservatism of this study's approach to valuing New Jersey's natural capital.

The estimated ecoservice value for forest land does not include two important services: 1) slowing stormwater runoff, thus reducing peak flows and decreasing the amount of stormwater storage capacity needed, and 2) removing pollutants like sulfur and nitrogen dioxides, carbon monoxide, ozone, and particulates from the air. Based on a non-peer-reviewed analysis, the value of these services from forests may total about \$9.0 billion and \$8.5 billion respectively in present value terms.

References

Costanza, R. et al. 1997. The Value of the World's Ecosystem Services and Natural Capital. Nature 387: 253-260.

Daily, G. (ed.). 1997. Nature's Services: Societal Dependence on Natural Ecosystems. Island Press, Washington, D.C.

United Nations Environment Programme (UNEP). Millennium Ecosystem Assessment. Available at http://www.maweb.org/en/index.aspx.

19

Economic Value of River Banks[1]

Ann L. Riley

Editor's Introduction: This paper compares the costs of treatment of urban waste water by two methods—allowing it to be clarified by river flows and treating the same in a water treatment plant. The paper shows that treatment of urban waste water can be done at a lower cost by letting it flow in rivers naturally. This shows that river banks have a large economic value. The practice of making highways along rivers assumes that river banks have no economic value other than the land involved. This requires reconsideration.

River banks are efficient at processing organic matter and sediments. Finer sediments are removed by reactions with soil. Bacteria and fungi living in stream bottoms consume nutrients brought by polluting discharges and convert them into beneficial

1. Adapted from "Putting a Price on Riparian Corridors as Water Treatment Facilities."

compounds. Plants in river banks absorb harmful chemicals. They also provide aesthetic, biodiversity and carbon sequestration services. These services provided by river banks make them more cost-effective as water treatment plants.

Flood protection embankments calculate values of property saved etc. from floods but do not calculated value of these services provided by river banks. Likewise highways and residential complexes built on river banks ignore the value of these services.

Introduction

The monetary value of natural riparian environments that provide water quality treatment functions by processing nutrients, storing sediment, moderating temperatures, and other services can be estimated by calculating the costs associated with the construction of brick and mortar water treatment plants built to achieve similar functions. A demonstration urban runoff treatment plant built by the City of Santa Monica provides similar services as a 4,000-5,000 lineal foot riparian corridor does, and has annualized costs of approximately $730,000 per year over a 20-year period taking into account the revenues the plant may accrue by selling its recycled water. The annualized costs for similar water quality treatment services provided by a natural riparian system were estimated by recording the costs of protective easements, fee acquisition, and restoration of riparian areas located in an urban context. A large, federally-funded, multi-objective flood control project with water quality benefits annualized over 20 years has about the same costs as the treatment plant. Urban stream restoration projects for 5,000 lineal feet of stream with riparian habitat can range in cost from $15,550 to $155,000 per year annualized over 20 years. Cost estimates of a range of restoration scenarios are given, including land acquisition and maintenance costs. While costs for all the alternative water treatment scenarios were given in 2005 dollar values and annualized over 20 years for purposes of comparison, most riparian restoration projects will be providing benefits over a 100 year period or in perpetuity, while the life span of the structural plant is expected to be 20 years or less.

Cost Comparisons: Man-made Systems- Naturally Occurring Systems

The first of its kind, state-of-the-art stormwater treatment plant located in Santa Monica, California gives us the opportunity to compare the performance and costs of a physical "brick and mortar" stormwater facility with a naturally occurring riparian woodland and their respective abilities to affect the quality of stormwater runoff. The Santa Monica Urban Runoff Recycling Facility (SMURRF) collects polluted runoff from the Los Angeles area and reclaims it sufficiently so that it can be re-used for landscape irrigation or dual plumbing systems. The plant came on line in February 2001 and is located near the Santa Monica pier. The building design involved a collaboration of engineers and artists. The plant features interesting architecture, art, and on-going visitor tours with public education about urban stormwater runoff, making this interesting, pioneering engineering facility an engaging tourist attraction.

This plant is intriguing for reasons other than its merits as one-of-a kind centralizedstormwater collection and treatment facility. The presence of a physical plant and the costs associated with its construction and operations and maintenance provide the first opportunity to compare the costs and effectiveness of a man-made stormwater treatment plant with the treatment capabilities and costs of a natural, functioning riparian system. Of course there may be no costs associated with the presence of a naturally existing riparian corridor, and in these cases we can better understand their value by examining the cost of protecting them. If we do allocate financial resources to protecting riparian resources or to restoring degraded waterways this gives us a basis for comparing the relative values and relative costs of these natural systems compared to building structural systems.

The Policy Context For The Comparisons

Ecologically functioning riparian environments are valued for the fact that they provide aquatic and terrestrial habitat for fish, amphibians, reptiles, mammals, and birds, and recreational and open space opportunities for the public. Yet little or no research appears to be available on the economic benefits of riparian areas to society for their water quality treatment functions. Riparian areas improve water quality by removing nutrients, improving dissolved oxygen, storing sediment and regulating temperatures among other benefits. These benefits can be achieved by protecting existing healthy riparian environments, or by restoring degraded areas into functioning ecosystems. Protection can be achieved by voluntary ecologically sound landowner practices, and/or through regulation, conservation easements, or fee purchase of riparian corridors. One method of assigning monetary value to these natural systems therefore, is to record what society pays to prevent farming or other land uses in these areas, pass protection regulations, purchase easements or full public rights to the riparian land, and/or to restore the ecosystem. In many circumstances, particularly in urban environments, the monetary costs of protecting a healthy system can be difficult to estimate. The comparison made here emphasizes the relative benefits and costs of two different types of stormwater treatment options which is typically a ubiquitous management problem in urban areas. For that reason, this paper focuses on the more readily available data on the costs and benefits of urban stream and riparian corridor restoration as one method of deriving their monetary value.

The SMURRF Plant Functions and Costs

The SMURRF Plant construction costs were approximately $13.6 million dollars (in $2005 dollars) (City of Santa Monica 2003), and the annual maintenance and operations costs are about $200,000 a year (Shapiro 2005). The plant treats about

320,000 gallons of runoff a day (City of Santa Monica 2003). One function of the plant is to remove fine sediments from the water, which is accomplished with a rotating drum screen. A second chamber removes grit and sand. Oil and grease are then removed in a unit that aerates the water using a compressed air unit (the dissolved air flotation unit.) This unit brings the oil and greases to the top so they can be skimmed off.

The next process in the plant is micro-filtration, which helps reduce the turbidity of the water by forcing the water through membranes. The membranes have to be periodically cleaned of pollutant build-up. The final step in the treatment process is to disinfect bacteria and viruses by passing the water under ultraviolet radiation lamps. The basic functions of the plant therefore are to filter sediment, reduce turbidity, trap oil and grease, and treat bacteria and viruses. Removal of sediment can also benefit removal of nutrients and other pollutants that may adhere to it (City of Santa Monica 2003). A separate trash collecting unit, which cost $200,000, catches trash from about 50,000 gallons a day before it enters the plant (Shapiro 2005).

Comparing A Treatment Plant To A Stream

In order to compare the costs and benefits of a plant against the costs and benefits of a riparian corridor we need to identify whether the water treatment functions of the plant and the riparian corridor are similar, and we need to compare the treatment of similar quantities of stormwater. The SMURRF plant treats approximately 320,000 gallons of water a day. The water treated is not wet weather runoff but dry weather run-off collected from about 5,100 urbanized acres. Stormwater flows from winter rainfall continue to run untreated into the ocean (Shapiro 2005). Theoretically, the plant could be expanded to treat wet and dry weather runoff, but for now I am going to assume that the costs per gallon of either dry season or wet season runoff are comparable. It is important to keep in mind that the plant may treat runoff from 5,100 acres but only treats a small portion of the runoff from that acreage. Therefore, we cannot use as a basis of comparison the number of acres served by our man made and natural "facilities," but we need to compare systems that can accommodate similar quantities of water.

A stream flowing at 1 cfs (cubic foot per second) produces a volume of water equal to 646,272 gallons per day. The 320,000 gallons treated by the plant equates to about .5 cfs flow per day. Using watershed and hydrologic information from a Bay Area stream we can estimate the size of the drainage area and creek that would produce a flow of about .5 cfs and then evaluate the ability of a stream of this scale to treat stormwater naturally. We can also compare the costs and benefits associated with restoring a length of stream that would treat a similar average annual flow to the costs and benefits of the stormwater plant.

The San Francisco Bay Area creek we will use for a costs and benefits comparison with the plant is Wildcat Creek located in the cities of Richmond and San Pablo, and the East Bay Regional Park system in Alameda and Contra Costa Counties.

The length of the Wildcat Creek stream channel is about 5,280 feet for each square mile of watershed drained, and the average daily flow from this square mile is about .64 cfs. The scale of this project and the discharges produced by this reach if considered separated from other watershed runoff, make it a reasonable case study with which to make comparisons to the SMURRF plant, which treats an average daily discharge of .5 cfs.

Natural Riparian Systems Functions and Costs

Research and collected field data is now available that addresses the issue of not only the water treatment functions riparian systems perform but also the area of the natural systems that produce the treatment results. A significant body of water quality research details the ability of riparian systems to store sediment, and retain and transform excess nutrients, pesticides, and toxic substances. (Meyer et al. 2003; Klapproth and Johnson

2000); Wenger 1999; Osborne and Kovacic 1993; Peterjohn and Cornell 1984; Chagrin River Watershed Partners 2006). The literature represents a wide range of environmental conditions and landscapes and therefore produces a range of quantifiable findings. For example, researchers in Corvalis, Oregon found that 60 to 80 percent of sediment generated from forest roads were captured by less than 250 feet of a healthy riparian system in point bars and pools, and their measurements indicated that stream systems could store sediment for as long as 114 years (Meyer et al. 2003). A study in the Southern Appalachian Mountains indicates that phosphorous- and nitrogen-containing compound ammonium traveled less than 65 feet downstream before being removed from the water by riparian areas (Meyer et al. 2003). First order headwater streams in the northeastern United States have been found to be responsible for 90 percent phosphorus removal (Meyer et al. 2003). A mathematical model based on research in 14 headwater streams throughout the country shows that 64 percent of inorganic nitrogen entering a small stream is transformed within 3,000 feet of stream channel (Naiman et al. 1997).

In general, riparian areas are found to be efficient at processing organic matter and sediments, and sediment bound pollutants carried in surface runoff are deposited effectively in riparian forests and floodplain areas. The finer sediments are removed from runoff as a result of deposition and erosion, infiltration, dilution, and adsorption/desorption reactions with woodland soil and litter (Bhowmilk et al. 1980). Riparian systems are known to have significant impacts on water temperatures and microclimates (Naiman et al.1997).

Research also indicates that healthy aquatic systems can transform animal waste and chemical fertilizers into less harmful substances. Vegetated buffers and protected riparian areas with contiguous riparian corridors have been shown to be effective in reducing pathogens such as coliform and cryptosporidium parvuum (Meyer et al. 2003; Tate, et al. 2004; Tate, III, 1978; Balance Hydrologics 2007).

The Wildcat Creek restoration project of 5,000 lineal feet, which has an average width of 60 feet, should have similar water quality treatment capacities as the SMURRF plant in respect to sediment removal, nutrient absorption, and breakdown of grease and oils, as described above. This riparian area also has the inherent capacity to reduce bacteria and viruses. The ultraviolet light treatment for pathogens is likely a more consistently reliable treatment for the latter; therefore, this may be the one area in which natural riparian system do not have equal treatment capacity.

If it is fair to argue that, conservatively, this 4,000 to 5000 foot riparian environment can perform water quality functions equivalent to this plant, can we then compare the relative cost of these systems?

Two of the most effective and commonly used methods to protect and or restore streams is to fence out livestock and or purchase conservation easements to remove riparian corridors from grazing or other agricultural uses.

To perform cost comparisons it is useful to annualize the cost of the SMURRF Plant over a twenty-year period using 2005-dollar values. The City Engineer's best estimate on the life of this plant is twenty years, based on the technology becoming obsolete by that time, although she cautions that breakdowns and replacements of machinery are inherent in the use of the new technology. The plant construction and land costs converted to 2005 dollars are $13.63 million and land values estimated by the county assessor are $33,000 (Higbee 2007). Annual maintenance costs are now approximately $200,000 per year. Therefore, the SMURRF construction and operations and maintenance costs annualized over this length of time come to $883,600 per year for the treatment of .5 cfs per day (Lowell 2005).

There are a number of scenarios for urban riparian protection and management that we can now compare against this dollar value. A wide range of restoration projects can be selected for making comparisons with, and it is a challenge to define a "typical" restoration project. The following restoration case used to make benefit and costs comparisons with the SMURRF project is a stream channel, floodplain, and riparian restoration project planned, designed, and constructed through a multi-agency watershed council. The project used for the comparisons involves a 250-foot-wide, 5,000-foot-long stream channel and its associated riparian habitat along Wildcat Creek. All values shown for costs use published data contained in government reports, which were then converted to 2005 dollar values based on the Consumer Price Index. The Wildcat Creek case allows us to compare very high and low range costs associated with stream restoration projects that occurred along the same reach of channel at different times.

Between 1986 and 1989, the Army Corps of Engineers, in partnership with Contra Costa County, constructed a multi-objective flood damage reduction project which included acquisition of the 250-foot-wide-corridor, and creation of a floodplain, vegetated corridor, and stream channel within the 250-foot-wide-corridor. Objectives of the project were to provide for a naturally functioning bankfull stream channel and adjacent floodplain, and protection of a riparian corridor. Army Corps projects represent the high end of costs for stream and river restoration work. The costs in this case equated to $2,706 per lineal foot. The total construction costs for 5,000 lineal feet (the total length of the project on Wildcat Creek was 10,000 feet) was $13.5 million, and land costs and relocation costs were $1.9 million for a total project cost of about $15.4 million. The annual maintenance costs expended by the county for this project area and staff support for the watershed council, which oversees the long tem management of the project area, is approximately $4,500 per year, for a twenty-year maintenance cost of $90,000. The annualized costs over a twenty year period of time for the high end cost Army Corps project is approximately $777,000 a year. For purposes of calculating a cost-benefit ratio for a multi-objective flood control project the Army Corps s amortizes the costs and benefits over a 100 year period).

In 2000, the watershed council implemented a 5,000-lineal-foot project along a reach in the same corridor to bring the project into conformance with the latest in geomorphic and engineering design knowledge and to provide a stream system with improved environmental values that could better maintain itself in an equilibrium condition. This project restored the stream channel to new dimensions, increased its sinuosity, and increased the average width of the riparian corridor from 30 to 55 feet. The 2000 project represents a major design and construction effort of the county and a non-profit organization; however, the project represents the lower end of the costs spectrum for restoration work at only $24.00 per lineal foot, for a total cost of $113,400. The Army Corps did provide a design document that helped validate the restoration design prepared by the non-profit organization. If the cost of that document is included, the cost of this restoration project is increased to $221,100, with a per lineal foot cost of $44.00. The annualized cost of this 2000 restoration project for a twenty-year period is $15,550 per year including maintenance costs; if we add in the original land acquisition costs covered by the earlier project the average annual cost increases to $112,000.

These costs provide actual figures for expensive and low cost projects; therefore, it is also useful to estimate costs that better represent average costs for stream restoration projects. Based on the experience of the author, who is involved in implementing stream restoration projects and comparing costs with other practitioners, a reasonable average lineal foot cost for a project of this scope conducted in 2005 in the San Francisco Bay Area would range between $300 and $600. Using the higher average value of $600 per lineal foot, a good estimate for a 5,000-foot riparian restoration project in the median range would be $3 million. Adding in average annual maintenance costs of $5,000 per year brings the annualized costs over a 20-year period to $155,000 per year.

Thus far, we have established that our total project cost comparisons on an annualized basis are $884,000 a year for the SMURRF plant, and the restoration projects range from $777,000 to $155,000 and $15,550 a year. Research indicates that the wide riparian and floodplain corridor and project length should be more than adequate to insure equivalent water treatment functions and benefits as the plant except psooibly virus control. The reason I choose to evaluate a 5,000 foot restoration corridor on Wildcat Creek is that this length of corridor, if it was viewed in isolation from the rest of the watershed, would produce approximately an equivalent average daily flow of about .6 cfs compared to the SMURRF plant average daily discharge of .5 cfs. However, we do have to recognize that we are probably not comparing equivalent water treatment functions because the average daily discharge that flows through this restored section of Wildcat Creek—because it is part of a larger watershed—is closer to 7 cfs, as opposed to the .5 cfs treated by the plant. Again it is reasonable to assume that the riparian corridor is affecting the quality of the total average daily 7 cfs. We could correct for the equivalent costs for "treatment" of .5 cfs by proportionately lowering the costs to approximate the costs per cfs treated. For example the treatment by a riparian system of 7 cfs average annual flow comes at a cost of $15,550, or $777,000. and therefore theoretically the costs for treating only .5-.6 cfs would recalculate the annualized costs at $1,100. or $55,000. for the expensive federal project.

Considering Multiple Benefits of Riparian Areas

This analysis so far restricts itself to only the comparable water treatment functions of the riparian system and the SMURRF plant. There are additional benefits of both the SMURRF Plant and the riparian systems that should be recognized and these can be described in either qualitative or quantitative terms. These benefits should also influence our evaluations of the relative values of these two very different systems.

The SMURRF plant also serves as a public education facility in which visitors can tour the plant and read interpretive displays about the plant and stormwater management. City records indicate that the plant averages about 230 visitors a year (Higbee 2005). Some of the water treated by the SMURRF plant is sold to customers, including the City of Santa Monica, for landscape irrigation and use in dual plumbing systems. Currently the water supplied by the plant is used in the new dual- plumbed Santa Monica Public Safety Building housing the police and fire departments, and the water is used to irrigate the grounds of the civic center parking structure, city parks, and cemetery, and Caltrans applies it to Santa Monica freeway landscaping. The income receipts for this water use currently total $32,000 a year based on 2003-2004 records (Lowell 2005). New water customers just now hooking up include a state-of-the art Rand Corporation Building and a commercial building know as The Water Gardens, which will be dual plumbed. It is estimated that this may increase the use of the water from the plant by 20 percent; therefore, receipts in the next few years could reasonably expect to increase to almost $40,000 annually. Unused flows return to the regional sewage treatment plant. It is very hard to predict future demand for the water cleaned by the plant because high volume estimates would be based on demand for newly constructed dual plumbing systems. The city water resources engineer's best estimate of a potential full use annual income if there is a demand for the full 230,000 gallons a day (based on a three tier pricing rate structure) is about $390,000 per year by 2016 (Lowell 2005). If we apply some optimistic assumptions about increasing demand over time for the water supply created by the plant, which includes a demand for the full amount treated by 2016, the plant will bring in an average annual income, based on a plant life of twenty years, of about $150,000 per year. (The plant may reasonably bring in total revenue of about $3,068,000 during its life span.) This benefit helps offset the annual management costs of $200,000, and lowers the total annualized costs of the plant to approximately $730,000 per year.

It can be argued that the SMURRF plant benefit of creating a recycled water supply cannot be compared against the benefits of the restoration project, which did not have the objective of developing a supply for purchase by urban water users. Likewise, it may not be appropriate to compare the substantial flood damage reduction benefits of the restoration projects against the benefits of the SMURRF plant because flood control was certainly not an intended objective of the SMURRF plant. In this case it is appropriate to compare water quality treatment functions only, and it is likewise appropriate to compare environmental and public education benefits, which are shared objectives of both projects. (The quantifiable flood damage reduction benefits are cited in the Projects Benefits Table.)

The restoration project has enabled an adjacent regional trail to be developed, and the project serves as a part of the educational opportunities for a very disadvantaged elementary school serving an impoverished community located next to the creek. The creek restoration area is also the focus for a Richmond High School environmental education program that serves about 25 students a year. The elementary school located next to the restored creek banks serves about 307 students a year. The project also serves as an anadromous steelhead (a threatened species) fisheries habitat restoration project and supports habitat and protection for the endangered California clapper rail and salt marsh harvest mouse. It is known that the restored riparian system offers habitat for mammals, raptors and other birds, and a range of aquatic organisms. One of the important objectives of the restoration project is to protect 200 acres of high quality brackish marsh from degradation by sedimentation. Environmental organizations hold regularly scheduled birding and wildlife hikes along the creek. The restored creek serves as the location for an on-going inner city youth environmental stewardship, training, and employment program that has involved an average of another 15 students on an annual basis for the past ten years, and there are varying numbers of community based water quality monitoring volunteers. This particular program has attracted over $200,000 in grants and donations to the community's desperately needed youth programs in a ten-year period.

The SMURRF plant serves as an educational facility for the general public and sponsors education tours for school classes. The plants sign-in records indicate an average of 230 visitors a year. The plant provides protection for recreationists at the beach in the Santa Monica pier area and a clean, safe area for use by surfers.

Economists can assign actual dollar values to the estimated visitor days and enjoyment of either the plant or the creek. They can also assign monetary values people place on seeing wildlife and protecting it. This paper is not intended to take the economic analysis to this level of detail. The point to be made here is that while the plant does have multiple human interest and educational values, there is a range of values intrinsic to a natural environment that a brick and mortar plant cannot emulate.

Analysis of Cost Comparisons

The costs data calculated compare three different categories of projects: a multi-objective flood damage reduction project that represents the high end of the costs spectrum; a local government-non-profit sponsored project that represents the low end of the spectrum; and average restoration costs based on the author's experience. The project benefits we are comparing are water quality and environmental education. The Army Corps flood control project did not explicitly have a water quality objective but was required by the regulatory process to avoid impacting water quality. The main purpose for which the Corps project was authorized was flood control, and this purpose or function alone produced annual net benefits. Because many of the costs are associated with delivering flood control benefits, the project does not provide a direct means of comparing the costs of two different kinds of water quality projects. However, what we do learn from this comparison is that the cost of a constructed stormwater treatment plant is equivalent to a moderate size Army Corps of Engineers flood control project.

Water Quality Program Policy Implications

Water quality programs have followed a logical progression from the first emphasis on the treatment of "point pollution" discharges from sewage treatment plants and industrial facilities. The second generation of water quality programs has focused on the avoidance and treatment of polluted runoff from "non-point" sources. The U.S. Environmental Protection Agency has identified six categories of non-point sources of polluted runoff including: urban properties and streets; farm fields, pastures and operations; forestry activities; marinas and recreational boating; hydromodifications of streams such as channelization, bank stabilization projects and stormwater discharge increases; and alteration of wetland and riparian areas. The three strategies applied to managing non-point sources pollution are prevention of pollution at the source, control and reduction of unavoidable runoff, and cleanup and remediation of pollutants that remain. Best management practices including environmentally sensitive land use and development site plans, and stormwater catchment and detention and filtering systems are common examples of source control and remediation. Protecting riparian areas, of course, directly addresses avoiding pollution from environmentally damaging hyrdomodifications and alterations of wetland areas. The evaluation most often missing from this non-point source management model is the recognition of the role of natural riparian areas to serve as part of the remediation system for runoff that escapes catchment and or detention near its source. This gives added value to riparian areas of not only addressing a part of the strategy to avoid degradation but also pro-actively remediating the impacts

Table 19.1. Comparison of Project Costs

*Wildcat Creek Project Costs**		*Costs of the SMURRF Plant*
A. 1986 U.S. Army Corps of Engineers multi-objective flood damage reduction project (This project was 10,000 feet long on Wildcat Creek, The costs shown here use the unit costs contained in the Design Memorandum applied to a 5,000-foot-long project.)	B. 2000 Watershed Council Restoration Project (executed by Contra Costa County and a non-profit) No land acquisition costs Channel excavation and partial revegetation for 5,021 feet of project channel	Values provided by City of Santa Monica.
Estimated construction costs for 5,000 lineal feet; $13,530,000.	Costs: $107,710 for design and construction by the watershed council	Construction costs: $13,638,740. Land Costs: The land used for construction of the SMURRF plant was in city ownership and is an odd shaped parcel, which made it infeasible to develop. The Los Angeles Assessors' office values the parcel of land, 2,783 sq. ft at $33,300.
Estimated permanent rights of way and relocation costs for 5,000 lineal feet: $1,916,424.	Construction Costs per lineal feet: $18.92	The footprint of the plant is 1,200 sq. ft. Average annual cost: $683,000.
Total project costs: $15,448,505.	$113,379 Army Corps planning	Average Annual Management costs: $200,000; for twenty years: $4 million (Plans are to offset

(Contd.)

(Contd Table 19.1)

*Wildcat Creek Project Costs**		*Costs of the SMURRF Plant*
Maintenance costs: $9,000 annually for 10,000 feet. Average annual costs for 5,000 feet are $4,500.	Total Cost: $221,089 Average Annual cost of restoration project: $11,054.00	maintenance costs by an estimated $3,068,000 income from recycled water sales over a 20-year period. This would reduce the total maintenance costs to $932,000 and average annual maintenance costs to $50,000.)
	Design and construction per lineal foot including Army Corps planning; $44.00 per foot	
	Average Annual Maintenance including management of the watershed council: $4,500.	
Total Annual Average Cost including lands and relocations cost ($1,916,424) and maintenance: $ 777,900.	**Total Average Annual Costs for 5,000-foot project including maintenance: $15,550.00**	**Total Average Annual Costs assuming projected income from water sales: $730,200.**

*Values shown are based on costs published in existing documents, converted to 2005 dollars based on the Consumer Price Index. Costs for all three projects amortized over 20 years.

Data from:

- Contra Costa County Wildcat and San Pablo Creeks Project Cost Summary.
- 1985 U.S. Army Corps of Engineers Design Memorandum for the Wildcat- San Pablo Creeks Flood Control Project, Richmond, Contra Costa County, Calif.
- City of Santa Monica.

Table 19.2. Comparison of Projects Benefits

Summary of Benefits of the Wildcat Creek Project for 100 years or less	*Summary of Benefits of the SMURRF Project for 20 years or less*
• 6.9 acres of high quality riparian corridor with a diversity of species and forest tiers to support wildlife habitat • 5,000 lineal feet of fish habitat and habitat for other aquatic species • Water quality functions: sediment collection and storage; nutrient uptake and conversion; bacteria reduction • Watershed Council conducts biannual community sponsored program of trash clean up	• 1,200 sq.ft. educational facility for the public. Visitors recorded averaged 230 annually • Partial trash collection • Treatment of low-flow dry weather runoff • Water Quality functions: sediment removal; nutrient removal to a water treatment plant for further treatment; bacterial treatment, and virus control • Protection of the Santa Monica beach and the surfers and other public who frequent the ocean in the area
• Water quality functions for average annual flows and greater magnitude flows • Flood storage and conveyance sufficient to protect the surrounding community from the damages associated with the one in one hundred year flood. Estimated average annual savings from avoided flood control damages calculated by the Army Corps of Engineers in 1986 for the period 1988-2088 is $871,000. • Active, hands-on environmental education experiences including water	• Income from the sale of recycled water averages $153,000 a year. • Water conservation for avoidance of use of equivalent potable supplies.

(Contd.)

(Contd. Table 19.2)

Summary of Benefits of the Wildcat Creek Project for 100 years or less	*Summary of Benefits of the SMURRF Project for 20 years or less*
quality monitoring, and cleanup and revegetation projects for 340 plus elementary school students and other local public schools and community members	
• Youth training and employment projects (ten year program attracted more than $200,000 to community youth programs)	
• Riparian corridor bird habitat and bird watching for hikers who use the creekside trail. (The Sierra Club, schools and other organizations sponsor hikes.)	
• Riparian corridor and floodplain protect 250 acres of downstream brackish and saltwater wetlands and San Francisco Bay water quality.	
• Endangered species habitat	

of various causes of non-point source pollution. A possible practical application of this information could be to assign water quality credits for meeting TMDL requirements in a watershed through the implementation of stream protection and restoration projects.

Current water quality budgets and priorities should evaluate the expenditures that have gone into treatment plants in the past and the expenditures that could occur in the future with mechanical stormwater treatment facilities, and use this evaluation as a budgeting framework for addressing the next generation of treatment systems. The comparisons described here indicate that the costs to restore degraded stream environments as fully functioning water treatment systems (which provide a significant range of other environmental benefits), can reasonably range from an average annual cost of $15,550. to a median of $155,000 and a high of $777,000 annualized over an inappropriately short period of twenty years and involving discharge amounts much greater than those addressed by a treatment plant. These costs and benefits compare to the annual cost of the stormwater treatment plant of around $730,000 for a system that treats a fraction of the amount of water and that has inherent limitations on additional environmental benefits. This represents a substantial magnitude in cost differences while the benefits of riparian environmental protection or restoration should be viewed as a more sustainable approach for attaining many more benefits through time.

References

Balance Hydrologics. 2007. Nitrate and Coliform Attenuation In Natural Area of Continuous Riparian Corridor. Memo Report on data collected for the Sanitary Survey for the City of Santa Cruz, California. Jan 17

Bhowmilk, N.G., A.P. Bonini, W.C.Bogner, R.P. Byrne. 1980. Hydraulics of Flow and Sediment Transport to the Kankee River in Illinois. Illinois State Water Survey Report of Investigation 98, Champaign, Ill.

Chagrin River Watershed Partners. 2006. Riparian setbacks, technical information for decision makers. Washington State.

City of Santa Monica. 2003. The Santa Monica Urban Runoff Recycling Facility, How The SMURRF Works.

Jean Higbee, Business Supervisor, Water Resources Davison, City of Santa Monica. 2005 and 2007. Personal communications.

Klapproth, Julia C., James E. Johnson. 2000. Understanding the Science Behind Riparian Forest Buffers: Effects on Water

Quality. Virginia Cooperative Extension, Publication No. 420-151. Virginia Polytechnic Institute and State University.

Susan Lowell, Water Resources Engineer, City of Santa Monica Water resources Division. 2005. Personal communication.

Meyer, Kaplan, Newbold, Strayer, et.al. 2003. *Where Rivers Are Born, The Scientific Imperative For Defending Small Streams and Wetlands*, published by American Rivers, Sierra Club Foundation.

Naiman, Robert J., Henri De'camps. 1997. The Ecology of Interfaces: Riparian Zones. *Annual Review, Ecological Systems* 28:621-58, Annual Reviews, Inc. Osborne L.L, D.A. Kovacic. 1993. Riparian Vegetated Buffer Strips in Water Quality Restoration and Stream Management. *Freshwater Biology* 29:243-258.

Peterjohn W.T., D.L. Cornell, 1984. Nutrient dynamics in an agricultural watershed: observations on the role of a riparian forest. *Ecology* 65:1466-1475.

Shapiro, Neal, Coordinator for Urban Runoff Programs, City of Santa Monica. 2005. Personal communication.

Tate, Kenneth, Maria Das Gracas, C. Pereira and Edward Atwill. 2004. Efficacy of Vegetated Buffer Strips for Retaining Cryptosporidum Parvuum. *Journal of Environmental Quality* 33:2243-2251, Madison Wisc.

Tate, Robert, III. 1978. Cultural and Environmental Factors Affecting the Longevity of Escherichia Coli in Histosols, *Applied and Environmental Microbiology* 925-929, May.

U.S. Geological Survey, Calendar Year Stream Flow Statistics for the Nation, Wildcat Creek, Vale Road at Richmond, Calif, Annual Flow Statistics.

Wenger, Seth. 1999. A Review of the Scientific Literature on Riparian Buffer Width, Extent, and Vegetation. Office of Public Service and Outreach, Institute of Ecology, University of Georgia, Athens.

Index

C

P

R

T

U

V

W